I0393889

Contents

Executive Summary

NATIONAL CLEAN DIESEL CAMPAIGN

REDUCING emissions from diesel engines is one of the most important public health challenges facing the country. Despite EPA's stringent diesel engine and fuel standards taking effect over the next decade, the 20 million engines already in use will continue to emit large amounts of nitrogen oxides (NO_x) and particulate matter (PM)—both of which will contribute to serious public health problems for years to come.

Fortunately, a variety of cost-effective technologies can dramatically reduce harmful emissions, save fuel, and help our nation meet its clean air and sustainability goals. To meet these challenges, the U.S. Environmental Protection Agency (EPA) established the National Clean Diesel Campaign (NCDC). NCDC consists of both regulatory programs to address new engines and innovative nonregulatory programs to address the millions of diesel engines already in use. EPA standards apply to *new* diesel engines, and because these engines can last a long time, solutions are needed to reduce harmful emissions from the existing fleet. These innovative approaches promote a variety of emission reduction strategies such as retrofitting, repairing, replacing, and repowering engines; reducing idling; and switching to cleaner fuels.

Through a dynamic network of Regional Collaboratives, whose development EPA initiated, environmental groups, industry, and government were inspired and motivated—despite their sometimes conflicting perspectives—to unite behind a common goal. NCDC mobilized diverse and unusual partners with historic differences to work together, creating broad support based on the urgency of the public health problem and bringing new technologies into use years earlier than would otherwise have occurred.

In 2008, Congress appropriated funding for the Diesel Emissions Reduction Act (DERA) Program under the Energy Policy Act of 2005 (EPAct 2005) to reduce emissions from heavy-duty diesel engines in the existing fleet. In the first year of the DERA program, EPA awarded $49.2 million for diesel emission reduction programs across the country.[1] EPA gave priority to projects that:

- Demonstrated a clear public health benefit in areas with high population density and poor air quality.
- Fostered cost-effective strategies that maximized the useful life of a certified engine configuration, verified technology, or emerging technology.
- Conserved diesel fuel.
- Used cleaner fuels.

---

[1] Congress also provided $10 million for grants for cost-effective emission reduction projects for two California air quality management districts (South Coast and San Joaquin).

The fiscal year (FY) 2008 funding enabled EPA to award 119 grants, which will lead to emission reductions of approximately 46,000 tons of NO_x and 2,200 tons of fine particulate matter ($PM_{2.5}$). The PM-related emission reductions translate into a significant quantifiable public health benefit of approximately $580 million to $1.4 billion. Additionally, projects funded under these grants will save 3.2 million gallons of fuel per year for a cost savings to operators of more than $8 million each year (at $2.50 per gallon). More than 14,000 diesel-powered vehicles and pieces of equipment are cleaner as a result of the first year of this program. More than two-thirds of these vehicles are trucks and school buses that were upgraded through EPA's sector outreach programs such as Clean School Bus USA.

DERA directs EPA to fund two different components: a national competition and a state allocation program. The national program, with 70 percent of the funding, consists of three separate competitions:

1. The National Clean Diesel Funding Assistance Program
2. The National Clean Diesel Emerging Technologies Program
3. The SmartWay Clean Diesel Finance Program

The State Clean Diesel Grant and Loan Program utilizes the remaining 30 percent of the funding.

National Clean Diesel Campaign Funding Structure

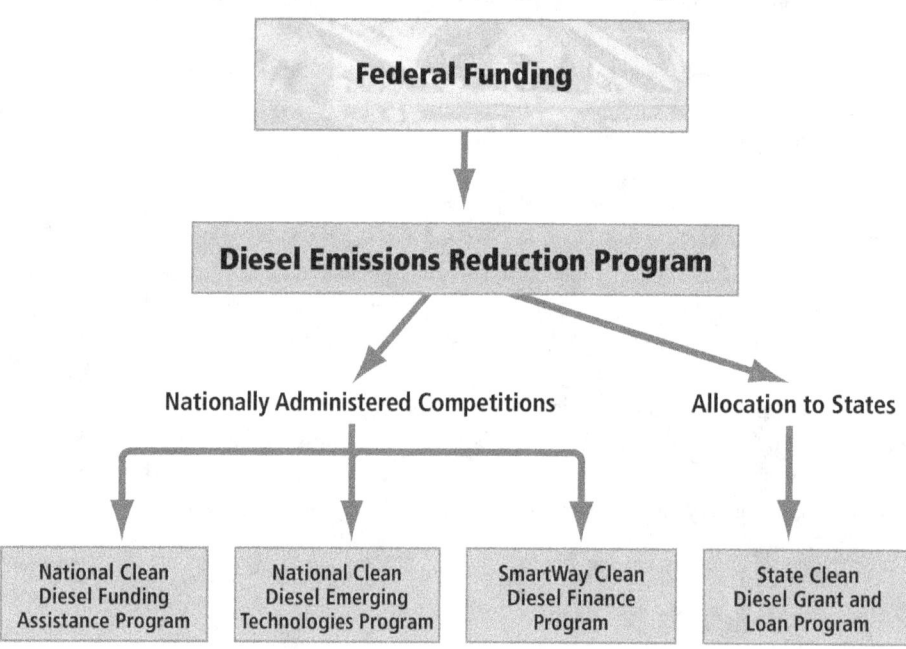

NCDC programs deploy proven technologies much earlier than would occur otherwise, accelerate the use of emerging technologies, and provide innovative financial incentives that make the business case for doing the right thing for the environment. States are key partners in the success of clean diesel programs and are given the opportunity through the state program to enhance their leadership on air quality issues.

NATIONAL CLEAN DIESEL PROGRAMS

National Clean Diesel Funding Assistance Program. The response to EPA's National Clean Diesel Funding Assistance Program was overwhelmingly positive. This program allows communities to meet their most pressing needs to reduce harmful diesel emissions. Communities benefit by activities as diverse as retrofitting school buses so that children's exposures are reduced, repowering locomotives used at seaports to save fuel as well as reduce emissions in neighborhoods, and replacing high-emitting construction equipment used to build hospitals and our nation's roads.

EPA's seven Regional Collaboratives held grant competitions. In FY 2008, applicants nationwide submitted 236 applications, requesting more than $144 million and offering approximately $81 million in matching funds. Demand for funding under this component of DERA exceeded available funds by 5:1. The variety of needs and approaches reflects the diversity of air quality challenges facing local communities across the country. After careful evaluation and ranking, EPA awarded approximately $30 million to 60 applicants, who provided matching funds of approximately $35 million. The high level of response to this program shows the desire communities have to clean up their fleets. The response also shows the ability of federal dollars to attract funds from other partners to accelerate the introduction of cost-effective clean diesel technologies.

National Clean Diesel Emerging Technologies Program. The emerging technology program, for which EPA may make available up to 10 percent of the funding, fosters the deployment of cutting-edge technologies and encourages private-sector investment in innovation. The program promotes the deployment of innovative technologies (those not yet verified or certified by EPA or the California Air Resources Board) by providing funding to develop and evaluate these technologies in the field. EPA received 10 eligible applications that requested $5 million and offered more than $1.1 million in matching funds. EPA ultimately awarded six grants totaling $3.7 million with $1.1 million provided as matching funds. These grants will demonstrate new technologies, including technologies that reduce NO_x. This program will help expand the currently limited retrofit options for nonroad engines, such as those used in construction equipment and on marine vessels.

SmartWay Clean Diesel Finance Program. For the first time, DERA gives EPA the authority to establish national low-cost revolving loan and other financing programs to provide funding to fleets to reduce diesel emissions. EPA exercised this authority by awarding grants where the recipients could provide innovative financing methods for clean diesel technologies. For the first time, partners are able to provide financial incentives (e.g., low-cost revolving loans, rebates, below-market rates) that stretch the federal dollar further. A small investment of federal funds leverages significant private investment and spurs the purchase of fuel-saving and emission reduction technologies for trucks. Under the SmartWay Clean

Diesel Finance Program competition, EPA received four grant applications requesting approximately $9.5 million that would leverage $44 million in outside funds, to establish innovative financing programs. EPA awarded three grants totaling approximately $3.4 million, leveraging $19 million. These grants establish national financing programs that provide funding to small- and medium-sized trucking companies to purchase clean diesel technologies. These trucking companies include many small businesses that would otherwise be unlikely to update their fleets with cleaner, more fuel-efficient equipment.

STATE CLEAN DIESEL GRANT AND LOAN PROGRAM

Recognizing that many states, such as California, Texas, Washington, Illinois, Maine, and New York, have developed successful clean diesel programs over the years, Congress included a state-only component in DERA to recognize the vital role played by states. In contrast to the national competitive program, the State Clean Diesel Grant and Loan Program is a formula allocation program. By meeting certain requirements, any state is eligible to participate. According to the statute, if adequate appropriations are available, 30 percent of any funds appropriated for DERA must be allocated to the states through this program.

In the first year of the state program, all 50 states elected to participate and established new clean diesel projects. The programs responded to state needs, focusing on a variety of sectors, such as school buses, construction, freight, refuse haulers, and transit buses. Based on funding availability and the states' ability to provide matching funds, individual state grant amounts ranged from $196,880 to $492,200. EPA funded the state programs with almost $13 million. Thirty-three states matched the federal dollars by providing more than $6.3 million in matching funds altogether.

COST-EFFECTIVE RESULTS OF PROVEN TECHNOLOGICAL SOLUTIONS

In this report, EPA analyzes the overall cost-effectiveness of various diesel emission reduction strategies. EPA compares the amount of federal grant funds used to the lifetime tons reduced to evaluate the effectiveness of federal investments in spurring diesel emission reductions. The Agency concludes that the innovative finance provisions are an important new approach with an impressive ability to leverage significant additional resources.

In the first year of funding, clean diesel strategies have proven to be about as cost-effective in reducing air pollution and health impacts associated with diesel emissions as other EPA programs, as shown in the table on page 7. Under the National Clean Diesel Funding Assistance Program, EPA calculated the average cost-effectiveness of federal investment to be approximately $27,700 per ton of PM reduced over the lifetime of the project and just under $2,000 per lifetime ton of NO_x. In the SmartWay Clean Diesel Finance Program, the cost-effectiveness was about $9,000 in federal funds per lifetime ton of PM and $400 in federal funds per lifetime ton of NO_x. The average cost of federal investment for the state program was approximately $16,700 per lifetime ton of PM reduced and $600 per lifetime ton of NO_x. These per ton costs compare very favorably with strategies used to attain national ambient air quality standards, such as stationary source standards that range from $1,000 to $20,000 and as high as $100,000 per ton of $PM_{2.5}$ on an annualized basis.[2, 3] However, DERA projects are on average[4] less cost-effective for PM than regulatory programs designed to set emissions standards for new diesel engines, such as the 2007 Heavy-Duty Highway diesel emissions standards. EPA has estimated that this regulation has a cost-effectiveness of $14,200 per ton PM and $2,100 per ton NO_x.

Average Cost-Effectiveness of DERA Projects

DERA Program	$PM_{2.5}$	NO_x
National Clean Diesel Funding Assistance Program	$27,700	$2,000
SmartWay Clean Diesel Finance Program	$9,000	$400
State Clean Diesel Grant and Loan Program	$16,700	$600

Regulatory Programs	$PM_{2.5}$	NO_x
Heavy-Duty Engine and Highway Diesel Fuel Requirements	$14,200	$2,100
PM NAAQS RIA	$3,000-15,000	–

Note 1: Values reported for the DERA program are for federal funds per ton pollutant reduced over project lifetime. Lifetime tons are not discounted. This report presents two different types of cost-effectiveness: federal funds per lifetime tons and total cost per ton. The comparisons for the regulatory programs are in annualized costs per ton.

Note 2: Values were rounded to nearest $100. Federal funds do not include matching amounts.

Note 3: The Emerging Technologies program is designed to demonstrate new technologies. At this phase in the development of these technologies, costs would be expected to be higher than when at a more fully mature commercial scale of deployment stage. Therefore, the cost comparisons with the other national program components have been omitted.

The first year of implementation for the Diesel Emissions Reduction Program has been extremely successful. By harnessing overwhelming support from environmental organizations, state and local governments, industry, technology vendors, and other groups, EPA has been able to magnify the available dollars and strategically fund important emission reduction and fuel-saving projects. In total, EPA awarded 119 grants in the first year of the DERA program. More than 14,000 vehicles and pieces of nonroad equipment in a wide array of sectors will be cleaner as a direct result of this program. New clean diesel programs have been established in every state, and technologies have been advanced for the future.

In addition, the American Recovery and Reinvestment Act of 2009 (Recovery Act) provided $300 million in new funding for the Diesel Emissions Reduction Program. EPA has already awarded more than $85 million through the State Clean Diesel Grant program, and those funds are currently at work creating green jobs and reducing air pollution. EPA is in the process of awarding national competitive grants under the Diesel Emissions Reduction Program with additional goals of creating and preserving jobs and promoting economic recovery. In response to this competitive program, EPA received more than 600 applications requesting approximately $2 billion and offering more than $2 billion in matching funds. This clearly demonstrates the high level of interest in clean diesel programs across the country. Implementation of the EPAct 2005's Diesel Emissions Reduction Program is a critical step toward reaching our national clean air goals and protecting public health and the environment for all Americans for generations to come.

---

[2] *National Ambient Air Quality Standards for Particle Pollution Regulatory Impact Assessment* (PM NAAQS RIA 10-06-06, Chapter 3, pg. 3-14, Table 3.2): Compares other stationary and area sources for PM reduction. Report is available at www.epa.gov/ttn/ecas/ria.html and www.epa.gov/ttn/ecas/regdata/RIAs/Chapter%203--Controls.pdf.

[3] *The Cost-Effectiveness of Heavy-Duty Diesel Retrofits and Other Mobile Source Emissions Reduction Projects and Programs* (EPA-420-B-07-006, May 2007, Appendix Table 3, pg. 13-15): Compares other mobile source programs cost-effectiveness for NO_x/VOCs. Report is available at www.epa.gov/otaq/stateresources/policy/general/420b07006.pdf.

[4] The weighted average federal cost of DERA projects is about $25,000 per ton PM and $1,500 per ton NO_x.

1. Introduction
Reducing Diesel Emissions Garners Broad Support

FROM the farm to the interstate highway to the neighborhood grocery store, we find diesel engines in every corner of society. Diesel engines power the movement of goods across the nation and help construct the buildings in which we live and work. Diesel engines aid in building the roads on which we travel and power the buses that carry millions of children to school safely each day. While diesel engines provide the mobility and power Americans require, are more fuel-efficient than gasoline engines, and serve as the workhorse of the nation's economy, exhaust from these engines sullies our skies with pollutants that harm our health and damage the environment.

Because of our reliance on diesel engines, reducing emissions from them is one of the most important public health challenges facing the country. Despite the U.S. Environmental Protection Agency's (EPA's) stringent diesel engine and fuel standards, which, for new engines beginning with engine model year 2007, are being phased in over the next decade, 20 million engines already in use continue to emit relatively large amounts of nitrogen oxides (NO_x) and fine particulate matter ($PM_{2.5}$). NO_x, which contributes to both ozone and particulate formation, as well as directly emitted $PM_{2.5}$, can lead to serious health conditions such as triggering asthma and worsening heart and lung disease. In addition, diesel engines emit black carbon, which might contribute to global climate change.

The problem of diesel emissions is not limited to a few discrete geographic areas or to one segment of the population. Harmful diesel emissions contribute to poor air quality in much of the country. No fewer than 141 million Americans—nearly half the population of the United States—live in areas that are designated as nonattainment with the eight-hour ozone standard and/or the $PM_{2.5}$ standard, based on the National Ambient Air Quality Standards (NAAQS).[5]

Due to the sheer numbers of engines in use and the volume of pollutants they emit, the health effects and environmental pollution stemming from diesel emissions could be substantial. Nationwide, in 2009 diesel emissions from mobile sources alone will account for approximately 300,000 tons of directly emitted $PM_{2.5}$ and 6.4 million tons of NO_x, which contribute to the formation of ozone and additional fine particles. These emissions will come from approximately 20 million engines operating in 2009, including approximately 13 million on-highway vehicles, 7 million nonroad engines, and 47,000 locomotive and marine engines. (See Figures 1 and 2.) Reducing diesel emissions quickly is vital to helping communities reach their public health goals.

---

[5] As of December 16, 2008, there are 57 eight-hour ozone nonattainment areas (1997) that consist of 293 full or partial counties, with a total population of almost 132 million. An additional 74 counties, where 16 million people reside, show air quality values that do not meet the 2008 ozone standard. As of December 16, 2008, there are 39 $PM_{2.5}$ nonattainment areas (1997) that consist of 208 full or partial counties, with a total population exceeding 88 million. On December 22, 2008, EPA designated nonattainment areas for the 2006 $PM_{2.5}$ NAAQS. As of December 22, 2008, there are 58 $PM_{2.5}$ nonattainment areas (2006) that consist of 211 full or partial counties.

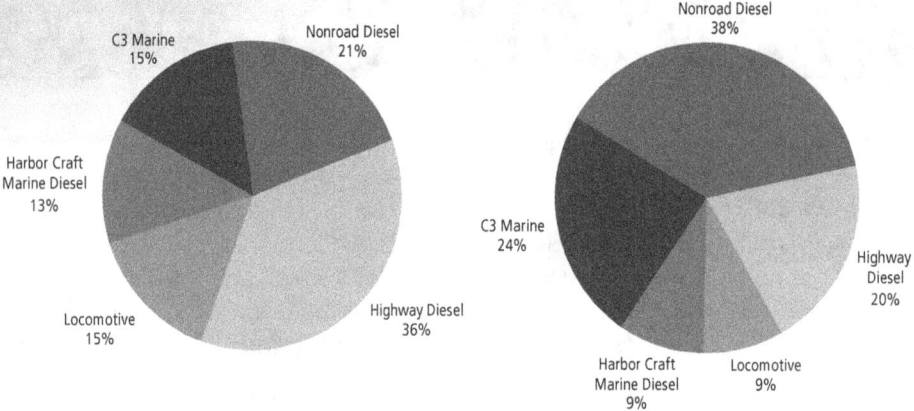

Figure 1. 2009 NO$_x$ Mobile Source Diesel Emissions for 50 States (6,400,000 tons)

- C3 Marine 15%
- Nonroad Diesel 21%
- Harbor Craft Marine Diesel 13%
- Locomotive 15%
- Highway Diesel 36%

Figure 2. 2009 Directly Emitted PM$_{2.5}$ Mobile Source Diesel Emissions for 50 States (300,000 tons)

- Nonroad Diesel 38%
- C3 Marine 24%
- Highway Diesel 20%
- Harbor Craft Marine Diesel 9%
- Locomotive 9%

Fortunately, a variety of cost-effective technologies can dramatically reduce harmful emissions, save fuel, and help our nation meet its clean air and sustainability goals. In 2000, to address the concerns of both new and existing diesel engines, EPA created the National Clean Diesel Campaign (NCDC), a partnership program that incorporates traditional regulatory approaches and innovative nonregulatory approaches to achieve results. The regulatory aspect of the program requires new engines and their fuels to meet stringent technology-based standards to reduce the amount of emissions released.

The innovative component of the program promotes the use of a variety of techniques to reduce emissions, including retrofitting, repairing, replacing, and repowering vehicles and equipment; reducing idling; and switching to cleaner fuels. To encourage these actions, NCDC cultivates the involvement of national, state, and local partners in the public and private sectors.

In 2008, for the first time ever, Congress appropriated funding under the Energy Policy Act of 2005 (EPAct 2005) to reduce emissions from diesel engines in the nation's existing fleet. In the first year of the program, the EPA's NCDC distributed $49.2 million to initiate diesel emission reduction projects and programs across the country. Reflecting the goals in the statute, EPA gave priority to projects that:

- Demonstrate a clear public health benefit and apply to areas with high population density and poor air quality.
- Use cost-effective strategies that maximize the useful life of a certified engine configuration, verified technology, or emerging technology.
- Conserve diesel fuel.
- Use cleaner fuels.

The EPAct of 2005, Title VII, Subtitle G, Section 794, requires that the EPA Administrator submit a report to Congress evaluating the implementation of the Diesel Emissions Reduction Act (DERA) Program. This report fulfills that requirement.

In the inaugural year of the program, EPA awarded 119 grants that are already achieving significant emission reductions, as detailed in Appendices A through F. As part of this effort, all 50 states have begun new state clean diesel program grants. The suite of national and state diesel emission reduction programs is a vital part of state and local efforts to reduce harmful air pollution. These first DERA grants are expected to significantly reduce emissions in communities across the country.

As shown in Table 1, the total emission reductions are noteworthy. The FY 2008 funding for the Diesel Emissions Reduction Program is estimated to reduce approximately 46,000 tons of NO_x and 2,200 tons of $PM_{2.5}$ over the life of the program. These emission reductions translate into a significant quantifiable public health benefit of approximately $580 million to $1.4 billion in quantifiable PM-related health benefits over the life of the program.[6]

In addition, the FY 2008 grants will save more than 3.2 million gallons of fuel per year, which equates to a savings in fuel costs to operators of more than $8 million per year (at $2.50 per gallon). This corresponds to 35,600 tons of carbon dioxide (CO_2) per year reduced. These fuel savings calculations only reflect the savings from idle reduction technologies and do not reflect the improved fuel economy that may result from engine replacements; thus the actual fuel savings might be higher than shown.

Diesel Exhaust Health Effects

Direct emissions from diesel engines, especially $PM_{2.5}$, NO_x, and sulfur oxides (SO_x), contribute to health problems. In addition, NO_x contributes to the formation of ozone and PM through chemical reactions.

$PM_{2.5}$ has been associated with an increased risk of premature mortality, increased hospital admissions for heart and lung disease, and increased respiratory symptoms. Long-term exposure to components of diesel exhaust, including diesel PM and diesel exhaust organic gases, are likely to pose a lung cancer hazard. Exposure to ozone can aggravate asthma and other respiratory diseases, leading to more asthma attacks, the use of additional medication, more severe symptoms that require a doctor's attention, more lost school and work days, more visits to the emergency room, increased hospitalizations, and even premature mortality. People in many areas of the United States experience short-term (one to three hours) and prolonged ozone exposures (six to eight hours), which have been linked to diminished lung function, greater respiratory symptoms, and increased hospital visits. Repeated exposure to ozone can increase susceptibility to respiratory infection and lung inflammation and can aggravate preexisting asthma. At sufficient concentrations, ozone can even cause permanent damage to the lungs, including the development of chronic respiratory illnesses. Children, outdoor workers, those who exercise outdoors, people with heart and lung disease, and the elderly are most at risk.

---

[6] Many of the benefits will accrue in the first five years of the program, and the benefits are discounted using a 3 percent rate.

Table 1. Estimated Air Emission Reductions of FY 2008 NCDC Programs Over Lifetime of Projects (Lifetime Tons)

Program	NO_x	PM	HC*	CO**	$CO_2^\dagger$
National Clean Diesel Funding Assistance Program	14,800	1,100	1,900	7,300	113,600
National Clean Diesel Emerging Technologies Program	600	30	30	100	—
SmartWay Clean Diesel Finance Program	9,600	400	500	1,600	238,200
State Clean Diesel Grant and Loan Program	21,000	800	2,000	5,700	112,900
Total	**46,000**	**2,200**	**4,400**	**14,700**	**464,600**

Note: Values were rounded to the nearest 100 lifetime tons or to one significant figure. Sums might not add due to rounding. Lifetime tons are not discounted.

* Hydrocarbon

** Carbon monoxide

$\dagger$ Does not include CO_2 reductions from engine replacements, repowers, or vehicle replacements.

The map in Figure 3 shows the location of the projects for the first year of the program. These projects are helping communities lower their public health risks from air pollution, accelerating the use of proven technologies, making engines cleaner and more efficient, encouraging new cutting-edge technologies, and utilizing new financial incentives to reach more fleets.

Figure 3. FY 2008 DERA Projects Across the Country

- National Program
- State Program

2. EPA's National Clean Diesel Campaign
A First-of-Its-Kind Campaign Lives Up to Its Goals

OVERVIEW

EPA's NCDC responds to the national need to reduce harmful diesel emissions. In April 2000, EPA began its Voluntary Diesel Retrofit Program with a few targeted projects. Using a small sample of the existing national fleet of school buses, delivery trucks, and transit buses, EPA demonstrated emission reduction technologies and cleaner fuels likely to be used for all new engines seven years before the 2007 effective date for the new engine standards.

Through this initial collaboration on retrofit projects, partners realized that they had found a way to put aside their sometimes-conflicting perspectives to unite behind a common goal. Fueled by this initial success, NCDC mobilized diverse partners with a wide range of perspectives—such as engine manufacturers, environmental groups, emission technology vendors, fuel suppliers, private fleet owners, state and local governments, and transportation officials—to work together, creating awareness of the urgency of the public health problem and accelerating the use of technologies years earlier than otherwise would have occurred. EPA provided seed money and technical expertise to create markets for new technologies and bring stakeholders together around an issue of mutual interest.

This tiny spark caught flame. Since 2000, EPA has awarded approximately 300 demonstration grants under the Clean Air Act for diverse sectors such as transit and school buses and marine, construction, and freight vehicles and equipment. These projects reduced unnecessary idling and promoted the use of alternative and cleaner fuels, retrofitting engines and replacing old engines or equipment with new, cleaner versions. Each year additional partners have joined NCDC, adding and showcasing new technologies. EPA tailored incentives to key sectors such as school buses, marine ports, construction, freight, transit, and agriculture. These successful demonstration projects resulted in the use of ultra-low-sulfur fuel in parts of 20 states well before the EPA mandate took effect. This cleaner diesel fuel directly reduces emissions from engines and enables the use of the most effective emission control devices. Now the use of this fuel is widespread and fleets are experiencing maintenance and emissions benefits of cleaner fuels.

VERIFYING THAT EMISSION REDUCTION TECHNOLOGIES WORK

Testing and evaluating the efficacy of diesel emission reduction technologies is a cornerstone of NCDC. From the beginning, fleet owners demanded to know that the technologies they wanted or were expected to use would work and would not harm their operations. State and local governments wanted assurance that they could rely on these strategies now and in the future to meet their air quality goals. In response to these needs, EPA created the Retrofit Technology Verification Program in 2002 to evaluate objectively the effectiveness of emission control technologies.

Through this program, EPA assures users that the actual emission benefits from retrofit technologies match those claimed by the manufacturer. The Agency also evaluates technologies that assist fleets in reducing idling. In addition, the California Air Resources Board (CARB) provides its own list of verified emission control technologies (www.arb.ca.gov/diesel/verdev/verdev.htm), which EPA recognizes. By verifying technologies, leading successful demonstrations, and operating recognition and education programs, these agencies have helped develop markets for new, greener technologies.

Yet this was only the beginning. Congress recognized the important role the verification process plays by making it a foundation of the implementation and integrity of the Diesel Emissions Reduction Program. Thus, in the national grant program, EPA *requires* the use of technologies that have been verified or certified by EPA or CARB when federal funds are applied. Congress also wanted to continue to spur innovation and encouraged the support and development of emerging technologies under DERA.

In establishing its verification program, EPA realized that verifying a technology only prior to its use is not enough. The technologies need "real world" testing to ensure that they maintain their performance throughout their useful life. As retrofit technologies are introduced into the market in higher volume, verifying the field performance and long-term durability of these products through in-use testing is increasingly essential. EPA requires all manufacturers to rigorously test each verified technology at two different stages during its useful life to confirm the percent of emissions reduction and evaluate durability. In-use testing offers confidence that verified retrofit technologies have been proven in the real world and will maintain that performance throughout their useful lives.

As a sign of the strength of the initial tests and of the technologies themselves, results from in-use testing are consistent with the original verified levels of emission reductions. In addition, EPA has been able to increase the level of reduction listed on its Web site for some of the technologies, given their outstanding performance in the in-use program.

To date, EPA and CARB have verified more than 50 emission control technologies. In addition, over the past seven years, EPA's SmartWay Transport Partnership has evaluated idle reduction technologies through grants, engineering analysis, and peer-reviewed reports. As of March 2009, EPA has designated six categories of idle reduction technologies as eligible for grant funding: 1) electrified parking spaces (truck stop electrification); 2) shore side power for ships, also known as cold ironing or Alternative Maritime Power; 3) auxiliary power units and generator sets; 4) fuel-operated heaters; 5) battery air-conditioning systems; and 6) thermal storage systems.

Clean Ports USA Helps Marine Ports Save Money and Decrease Emissions

Due to EPA's Clean Ports USA program (part of the National Clean Diesel Campaign), leading marine ports are saving fuel, increasing operational efficiency, and decreasing diesel emissions. In addition, several prominent ports have developed clean air action plans, which include greenhouse gas reduction goals, to assist them in reducing air pollution in gateway communities.

CLEANPORTSUSA

In January 2005, EPA partnered with the American Association of Port Authorities to create Clean Ports USA as an outreach- and incentive-based program designed to reduce emissions from existing diesel engines operated at marine ports. EPA is working with port authorities; terminal operators; and shipping, trucking and rail companies to promote environmentally superior technologies through education, incentives, and financial assistance. For more information, visit: www.epa.gov/cleandiesel/ports.

SmartWay Transport Partnership

EPA developed the SmartWay Transport Partnership Program to address the environmental impact of the freight system in the United States. The Smart-Way program is a voluntary public–private initiative designed to improve the environmental performance

of the freight delivery system through money-saving, market-based approaches. The goal of the program is to reduce emissions by promoting cost-effective strategies that reduce fuel consumption and air pollution, using such methods as eliminating unnecessary idling, installing emission control devices, and improving freight logistics.

The SmartWay program works hand in hand with the National Clean Diesel Campaign to promote diesel emission reduction strategies. The SmartWay program has pioneered innovative approaches to emission reduction, such as the creation of SmartWay-approved trucks to identify trucks with superior environmental performance and the development of low-cost financing for the purchase of fuel saving and emission control devices. The innovative financial strategies offer increased access to lenders who can provide financing for SmartWay-verified technologies and provide truck owners with low-cost financing for the purchase of SmartWay Upgrade Kits (which include idle reduction technologies, advanced aerodynamic technologies, and low-rolling resistance tires) or the purchase of cleaner used diesel trucks with aftertreatment devices. The SmartWay program also recognizes partners who achieve superior environmental performance. For more information, visit: www.epa.gov/smartway.

TAPPING INTO REGIONAL COLLABORATIVES

With the foundation of cost-effective, reliable technologies, EPA's partners wanted to reach more fleets in more sectors to reduce harmful diesel emissions. As a next step, beginning in 2003, EPA's regions began organizing Regional Clean Diesel Collaboratives with states, local nonprofit organizations, private industry, and municipalities. (See Figure 4 for a map.) By tying into a network of regional stakeholders, this collaborative structure is well-suited for achieving significant emission reductions across large geographic areas. Members of these collaboratives have agreed to collectively leverage additional funds and take a local approach to diesel emission mitigation. These collaboratives play a vital role in EPA's proactive, incentive-based approach to achieving superior environmental results. Information on the collaboratives can be found in Appendix F.

Figure 4. Regional Clean Diesel Collaboratives

REGULATING DIESEL ENGINES AND FUELS

While cost-effective technology and strong coalitions of committed stakeholders form the backbone of NCDC, another important component is a series of regulations designed to reduce pollution emitted from new diesel engines and their fuels.

Under the Clean Air Act, EPA developed and implements a suite of stringent regulations to mitigate emissions from new diesel engines and their fuels. These regulations apply to a variety of new engines, including those in trucks, buses, construction equipment, locomotives, harbor craft, and large ships. Over the long term, these standards will yield enormous public health and environmental benefits.

When fully implemented, these regulatory programs will reduce NO_x by about 7 million tons per year, $PM_{2.5}$ by more than 300,000 tons, and SO_x by about 800,000 tons. By 2030, when the regulatory programs are fully phased in and implemented, the net public health benefits total approximately $186 billion per year, including annually preventing 26,000 premature deaths, approximately 20,000 hospitalizations, and more than 3.3 million days lost from work due to respiratory problems as described in EPA's Regulatory Impact Assessments.

In general, these rules require stringent standards for PM and NO_x, based on levels achievable with advanced diesel aftertreatment technologies, such as diesel particulate filters (DPFs) for PM and lean NO_x traps or selective catalytic reduction (SCR) for NO_x. Additionally, each of these rules ultimately requires reducing fuel sulfur levels to 15 parts per million (ppm) to enable the use of advanced aftertreatment technologies.

NCDC's goal is to bring these impressive benefits to the American people earlier by accelerating the adoption of proven technologies through its strategic regulatory approach.

3. The Energy Policy Act of 2005, Diesel Emissions Reduction Act Program
Congressional Support Buoys Programs Nationwide

BACKGROUND

THE EPAct of 2005 provides EPA new grant and loan authority to promote diesel emission reductions and authorizes appropriations of up to $200 million per year to the Agency under the Diesel Emissions Reduction Program provisions (for FY 2007 through FY 2011). Congress appropriated funds under this statute for the first time in FY 2008, in the amount of $49.2 million. This portion of the report highlights key components of the statute and explains how EPA is implementing the program with state partners pursuant to Congress's direction in the statute.

DERA includes a national program and a state allocation program. The national program includes 70 percent of appropriated funds, and the state program makes available 30 percent. Under this act, EPA can offer competitive grants and, for the first time, low-cost revolving loans to eligible organizations and entities on a competitive basis.

To spur innovation, DERA allows for up to 10 percent of the national funds to be spent on "emerging technologies." Additionally, not less than 50 percent of the funds available for the national program must be used for the benefit of public fleets. DERA requires that engine configurations and technologies supported by the national program be verified or certified by EPA or CARB.

While the national program offers competitive grants in three categories, the state program is a single allocation program, which means that base funding is distributed to states using a specific formula based on participation, and incentive funding is also available for any states that match their base funding. Funds not claimed under this program—either if states decline to participate and/or do not match the base funding—revert to the national program. By meeting certain requirements, all of the 50 states and the District of Columbia are eligible to participate.[7]

Under the statute, EPA provides guidance to the states about the process for applications, permissible use of funds, and the cost-effectiveness of various emission reduction technologies. EPA established an annual deadline for submitting applications, an approval process, and a streamlined renewal process. Discretion is given to the governors for apportioning funds between grants and loans in each state. EPA published a *Federal Register* notice (73 FR 12728, March 10, 2008) that discussed how it intended to implement the state programs.

---

[7] The District of Columbia is now an eligible entity for the State Clean Diesel Grant and Loan Program, but it was not eligible when EPA awarded the FY 2008 state grants.

The Energy Policy Act of 2005

Under the EPAct 2005, EPA gives priority to those diesel emission reduction programs that:

1. Maximize public health benefits.

2. Are the most cost-effective.

3. Are in areas with high population density and poor air quality (including nonattainment areas or areas that require maintenance of national ambient air quality standards for a criteria pollutant, federal Class I areas, or areas with toxic air pollutant concerns).

4. Are in areas that receive a disproportionate quantity of air pollution from diesel fleets, including truck stops, ports, rail yards, terminals, and distribution centers, or that use a community-based multi-stakeholder collaborative process to reduce toxic emissions.

5. Include a certified engine configuration or verified technology that has a long expected useful life.

6. Maximize the useful life of any certified engine configuration or verified technology used or funded by the eligible entity.

7. Conserve diesel fuel.

8. Utilize ultra-low sulfur diesel fuel (15 ppm of sulfur content) ahead of EPA's mandate (for nonroad or marine projects).

Relevant sections of the EPAct 2005 that govern the DERA program are found in Title VII, Subtitle G.

Section 791, "Definitions"

Section 792, "National Grant and Loan Programs"

Section 793, "State Grant and Loan Program"

Section 794, "Evaluation and Report"

Section 795, "Outreach and Incentives"

Section 796, "Relationship to Clean Air Act"

Section 797, "Authorization for Appropriated Funds (2007–2011)"

With the foundations of its successful Clean Air Act demonstration programs, verification program, sector-based focus, Regional Collaboratives, and SmartWay brand, NCDC was well-poised to deploy cost-effective diesel emission reduction strategies even more broadly under DERA.

EPA's goal for the first year of DERA funding was to establish diesel emission reduction programs in every state and accelerate the adoption of clean diesel technologies in the existing fleet. The EPAct 2005 statute establishes a general framework for the Diesel Emissions Reduction Program, charging EPA to develop other details for how the total program is implemented. Ultimately, EPA designed four programs:

1. The National Clean Diesel Funding Assistance Program
2. The National Clean Diesel Emerging Technologies Program
3. The SmartWay Clean Diesel Finance Program
4. The State Clean Diesel Grant and Loan Program

These programs deploy proven technologies much earlier than would otherwise occur to meet critical local air quality needs. They accelerate the use of emerging technologies and provide innovative financial incentives that make the business case for reducing diesel emissions. Furthermore, states are key partners in the success of clean diesel programs, and the state program enhances their leadership in reducing harmful diesel emissions.

Figure 5. National Clean Diesel Campaign Funding Structure

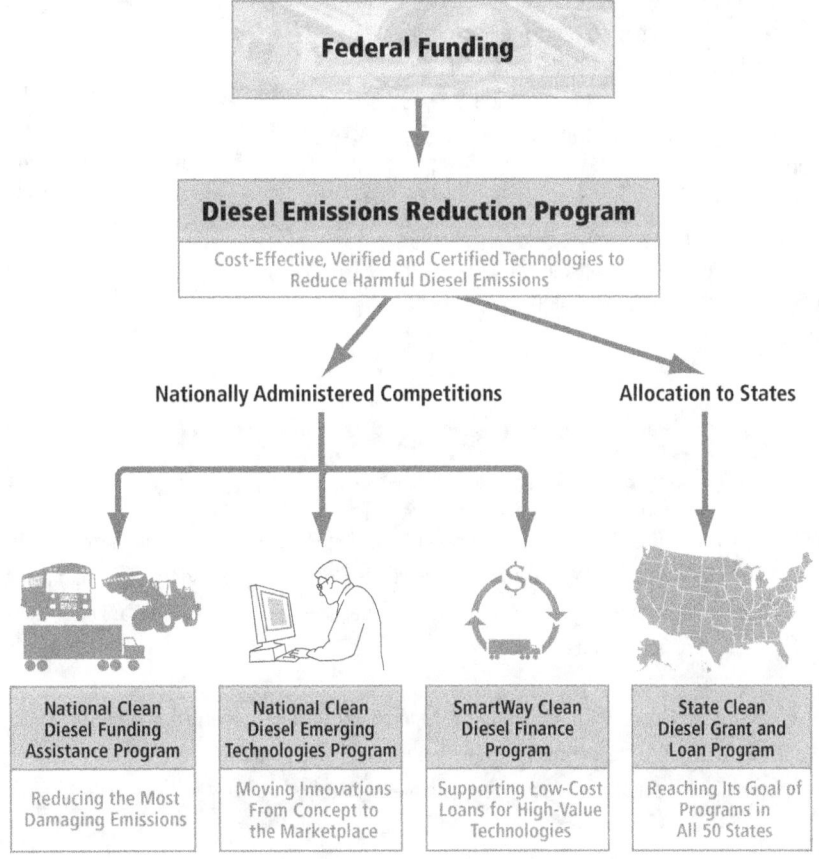

THE NATIONAL CLEAN DIESEL FUNDING ASSISTANCE PROGRAM
Reducing the Most Damaging Emissions

EPA received an overwhelmingly positive response to its Requests for Proposals (RFPs) under the National Clean Diesel Funding Assistance Program. This program offered the majority of the funding available, through grant competitions handled by the Agency's seven Regional Collaboratives. The Agency received 236 applications nationwide, requesting more than $144 million and offering $81 million in matching funds. Thus, applicants requested $5 for every $1 available.

Applicants requested funding to address their most pressing needs, such as retrofitting school buses so that children's exposures are reduced on their way to school, repowering locomotives used at seaports to save fuel and reduce emissions in neighborhoods while still delivering freight used in every part of commerce, and replacing high-emitting construction equipment used to build hospitals and our nation's roads.

After screening for eligibility and carefully ranking the applications, EPA selected 60 recipients across the country in the fall and winter of 2008—within nine months of the program start date—and awarded $29,039,803 in federal dollars, attracting approximately $35 million in matching funds. The federal share included $1,476,600 from unclaimed state funds (states that did not match their base funding). Appendix E lists the ranking criteria, reflecting the priorities listed on page 20. Table 2 summarizes the awards by region, and Appendix A contains a comprehensive list of the awardees and grant amounts.

Emissions Reduced From Nonroad Cargo Handling Equipment at Pacific Northwest Ports

Through its Clean Ports USA Program, EPA awarded the Puget Sound Clean Air Agency an $850,000 grant to fund replacements and retrofits, such as diesel particulate filters, of nonroad truck engines for cargo-handling equipment at Puget Sound ports.

This grant helps the port authorities achieve the goals in their Northwest Ports Clean Air Plan. Additionally, the Ports of Seattle and Tacoma and the Puget Sound Clean Air Agency are providing matching funds totaling $318,000.

Illinois School Buses Reap the Benefits of DERA Grants

EPA's Clean School Bus USA grants to Illinois EPA's Green Fleets Program, including school buses, totaled $678,600, plus a $654,700 match from Illinois EPA, funding a variety of clean diesel technologies, including diesel oxidation catalysts, diesel particulate filters, closed crankcase ventilation, and auxiliary power units.

The Illinois Green Fleets Program formally recognizes businesses, governmental entities, and organizations that use clean, domestic, renewable fuel vehicles in their fleet. The program also recognizes Illinois vehicle fleets equipped with diesel retrofits.

Table 2. FY 2008 Awards for the National Clean Diesel Funding Assistance Program by EPA Region

Region	Collaborative	Number of Applications	Number of Awarded Grants	Dollars Awarded
1	Northeast	16	6	$1,970,484
2	Northeast	33	9	$3,268,402
3	Mid-Atlantic	24	6	$3,277,166
4	Southeast	39	7	$3,789,289
5	Midwest	38	13	$4,879,049
6	Blue Skyways	20	3	$2,719,880
7	Blue Skyways	10	3	$1,986,802
8	Rocky Mountain	11	5	$1,755,645
9	West Coast	30	5	$3,953,079
10	West Coast	15	3	$1,440,007
Total		236	60	$29,039,803

Public Fleets Directly Benefit From DERA National Program Funds

Public fleets across the nation—including transit buses, school buses, refuse haulers, municipal trucks, snowplows, and fire trucks—are emitting less pollution in towns and cities as a direct result of the Diesel Emissions Reduction Program. In FY 2008, the Agency awarded 74 percent of competitively awarded national funds to public fleets. Section 792 (b)(2) of the EPAct 2005 requires that no less than 50 percent of the funds for the national program be spent for the benefit of public fleets.

Many partners stepped forward to deploy clean diesel technologies in public fleets to lead by example. Making the public-sector vehicles that do the nation's work the cleanest they can be creates a powerful example for other fleets to imitate.

THE NATIONAL CLEAN DIESEL EMERGING TECHNOLOGIES PROGRAM
Moving Innovations From Concept to Marketplace

EPA's NCDC has been demonstrating new technologies and creating new demands for environmentally superior products since its inception. To continue to encourage new technologies to meet the needs of fleets and air quality planners, Congress recognized the need for the DERA program to promote cutting-edge advanced technologies. EPA developed the National Clean Diesel Emerging Technologies Program to respond to Section 792 (b)(3)(B)(i), which directs that "...the Administrator shall provide not more than 10 percent of funds available for a fiscal year under this provision to eligible entities for the development and commercialization of emerging technologies."

In FY 2008, under the National Diesel Emerging Technologies Program, 10 applicants requested $5 million and offered more than $1 million in matching funds. EPA awarded six grants in five states for a total of $3.7 million in federal funds, with $1.1 million in matching funds. These funds supported NO_x reduction technologies such as SCR systems for on-highway trucks and nonroad construction equipment, greater emission controls from a combination of diesel oxidation catalysts (DOCs)/DPFs and exhaust gas recirculation (EGR) systems, and new techniques for harbor craft, such as an emission upgrade kit for two types of marine engines. The criteria by which the applicants were evaluated are presented in Appendix E. The new federal dollars used for these emerging technologies are vital to spurring investment by NCDC's partners.

The National Clean Diesel Emerging Technologies Program is an opportunity to advance new, cutting-edge technologies to reduce diesel emissions from the existing fleet. Building on EPA's verification program, EPA developed a process by which a manufacturer can request that its technology qualify in advance as an emerging technology. Through its collaborative network, EPA has been encouraging new technologies to reduce NO_x, save fuel, and perform better on a wider array of conditions and vehicles, including nonroad equipment and marine vessels.

To qualify as an emerging technology, manufacturers should be in the initial stages of the verification process with EPA or CARB and listed on EPA's Emerging Technology List. The manufacturer must also provide an approvable application for verification (including an explanation of the engineering principles of the technology and why the technology should be considered an emerging technology) and an approvable test plan. An emerging technology must be close to being, if not already, commercially available.

Emerging Technology

An emerging technology is defined in Section 791 of the Energy Policy Act to be "... a technology that is not certified or verified by the Administrator or the California Air Resources Board but for which an approvable application and test plan has been submitted for verification to the Administrator or the California Air Resources Board."

Once approved, emerging technologies are listed on EPA's Web site. Technologies may remain on EPA's Emerging Technology List for up to two years. During that two-year period, EPA estimates that manufacturers will be able to complete the necessary steps to obtain full verification. Once a technology is verified or certified, it will no longer be considered "emerging" but will then be fully eligible for projects under the National Clean Diesel Funding Assistance Program and then deployed more widely. This program will accelerate the pace of introduction of new diesel emission reduction strategies and will allow the DERA program to evolve to better serve the public.

EPA awarded a grant to the South Coast Metropolitan Air Quality Management District, in California, home to approximately 16 million people, to install selective catalytic reduction technology (SCRT) on 1999 to 2002 heavy-duty Class 8 on-road diesel trucks, which will reduce PM, NO_x, CO, and HC emissions by at least 90, 65, 85, and 90 percent, respectively. This project will improve the air quality of a high-population-density nonattainment area. At the same time, the project will produce useful data about the effectiveness and durability of this emerging technology.

THE SMARTWAY CLEAN DIESEL FINANCE PROGRAM
Supporting Low-Cost Loans for High-Value Technologies

Under the SmartWay Clean Diesel Finance Program, EPA received four applications requesting $9.5 million and leveraging $44 million. Ultimately, in September 2008, EPA awarded three competitive grants totaling $3.4 million to establish a national financing program that provides funding to small- and medium-sized trucking companies to purchase clean diesel technologies, including leveraging more than $19 million in additional funding.

These innovative financing programs will help smaller trucking firms lower their emissions, fuel costs, and carbon footprint by purchasing cleaner used trucks equipped with diesel exhaust filters and/or fuel-saving technologies. The program will also support idling and emission reduction technologies for 2,460 trucks. Appendix E presents the criteria by which the applicants were evaluated. Appendix C provides additional details about the projects.

DERA provided EPA with new authority to establish national grant and low-cost revolving loan programs to finance clean diesel projects. Administered by EPA's SmartWay Transport Partnership, the FY 2008 RFP for this program focused on establishing low-cost loan programs for retrofitting used pre-2007 highway vehicles and new or used pieces of nonroad equipment with EPA- or CARB-verified emission control and idle reduction technologies. EPA encouraged financing proposals to include strategies such as loan guarantees, equity investments that leverage additional funds, tax-exempt or taxable bonds to create a low-cost loan program, and revolving loan funds.

Using up to 10 percent of the funds available for the national program, the SmartWay Clean Diesel Finance Program is designed to allow eligible entities to create financing mechanisms to assist fleets with the purchase of EPA-approved fuel-saving and emission reduction technologies. To date, EPA has offered grantees the opportunity to develop innovative financing projects instead of issuing direct federal loans.

Loan programs are important because they enhance financial sustainability and allow the funds to be used multiple times as the loans are repaid and additional loans can be offered to purchasers of cleaner vehicles or technologies. They are able to leverage substantial nonfederal funds to increase the federal dollar's effectiveness. For example, in the case of a finance program using an equity investment, the award funds could be used to leverage additional funds from a lending institution to increase the amount available for loans.

Although potentially far reaching, this program presents unique challenges that are not faced in the National Clean Diesel Funding Assistance Program. Unlike public school bus fleets, where the fleet owners have a strong motivation to reduce emissions, most private diesel truck and equipment owners do not have the same incentives and may not be as receptive to installing diesel emission reduction technologies. EPA has attempted, in previous demonstration grants, to provide retrofit technologies to truck operators and has received only minimal interest.

To help manage these challenges, EPA's SmartWay program worked with a variety of stakeholders in the trucking industry including truckers, truck dealers, truck finance companies, and trucking associations. EPA concluded that the best incentive to install retrofit technologies on older trucks is to provide lower cost financing to purchase newer trucks already equipped with these technologies. In fact, the assessment shows that truckers will purchase cleaner diesel trucks equipped with advanced air pollution control devices if the monthly finance charge is $100 to $200 less than traditional trucks without these technologies. The federal grant funds can be used to lower the monthly finance cost by either extending the terms of the loans or reducing the interest rates.

Grant Offers Low-Interest Loans for SmartWay Upgrades

Using new authority to issue grants for loans and other innovative financing mechanisms, EPA awarded $1.13 million to the Community Development Transportation Lending Services, Inc. to establish a revolving loan fund for the purchase of used trucks that have SmartWay upgrades or will receive SmartWay upgrade kits.

The program uses the grant to provide a lower interest rate than borrowers would otherwise be able to get—in this case, 5.5 to 8.5 percent, based on each applicant's business history, available collateral, cash flow, credit score, and other factors. The loan payback period is three to six years.

THE STATE CLEAN DIESEL GRANT AND LOAN PROGRAM
Reaching Its Goal of Programs in All 50 States

As a telling indication of the need for and interest in diesel emission reductions throughout the United States, all 50 states elected to participate in the FY 2008 State Clean Diesel Grant and Loan Program. EPA apportioned a base amount of $196,880 to each state.[8] In addition, as another sign of the nationwide support for diesel emission reduction programs, 32 states were able to match the base funding, providing more than $6.3 million, and therefore received additional incentive funding, as allowed by DERA. For these states, EPA established a total grant of $492,200 (with $295,320 in federal funds and at least $196,880 in state funds).

With this money, each state established a clean diesel program within its air agency and worked in consultation with EPA to develop an effective work plan and application. The projects varied and utilized a number of emission reduction strategies. A total of 21 states concentrated their efforts on school buses; nine focused on the construction sector; and other states focused on agriculture, transit buses, marine ports, rail, or trucking. Several states that have never initiated a state clean diesel program before are starting new programs. Appendix D presents a list of projects.

EPA issued a *Federal Register* notice (73 FR 12728; March 10, 2008) that announced the funding opportunity and explained the main provisions of the State Clean Diesel Grant and Loan Program. The state program is not a competition but, rather, an allocation process in which the states indicate their interest in participating and applying for the funds, and EPA awards a specific allocation based on the total number of states whose applications are approved. Any state wishing to participate receives funding if the state submits an approvable work plan. Early on, the EPA Administrator sent a letter to the governor of each state outlining the goals of the State Clean Diesel Grant and Loan Program.

Provisions for State Uses of State Clean Diesel Grant and Loan Program Funds

- States had significant flexibility in their proposed work plan.

- States were encouraged to use funds for EPA- and/or CARB-verified or certified technologies, although there were cases where emerging technologies or other cost-effective strategies were allowed, as they advanced the goals of the program.

- States could use funds to select various state fleets or projects for funding.

- States were encouraged not to use funds for emission reductions that are mandated under federal, state, or local law (similar to Section 792 (d)(2), "Use of Funds—Regulatory Programs").

- States could use up to 15 percent for administrative costs.

---

[8] The District of Columbia is now an eligible entity for the State Clean Diesel Grant and Loan program, but it was not eligible when EPA awarded the FY 2008 state grants.

Inaugural Year Results in Impressive, Cost-Effective Reductions Across All Sectors

Receiving significant levels of funding changes the complexion of possibilities for communities struggling with ways to reduce harmful diesel pollution. Instead of small demonstrations, entire fleets can take advantage of proven technologies. Federal investments can attract private capital through low-cost loans under the new DERA provisions. Financially strapped states can move forward to maximize public health programs and improve air quality for their citizens.

In the first year of DERA funding, EPA's programs achieved what they set out to do. More than 14,000 diesel-powered vehicles and pieces of nonroad equipment became cleaner as an immediate and direct result of this work.

DERA Funding Supported Reductions Across All Sectors. As shown in Figure 6, the FY 2008 grants supported a wide array of projects in numerous sectors, from agriculture and construction to marine and seaports to rail to transit. Collectively, the power of many small actions across the range of commercial, municipal, and private endeavors yielded significant emission reductions.

State of Delaware Expands Its Program to Cover More Municipal Vehicles

EPA awarded the Delaware Department of Natural Resources and Environmental Control with a state clean diesel grant to retrofit 1998 Class 7 refuse haulers. Because the state put some of its own resources toward the program, EPA allocated additional incentive funding as well. This project is an excellent example of how the DERA funds allocated toward state programs can benefit the air quality of each state that opts into the program.

Figure 6. FY 2008 National Grants—Types and Numbers of Vehicles and Equipment

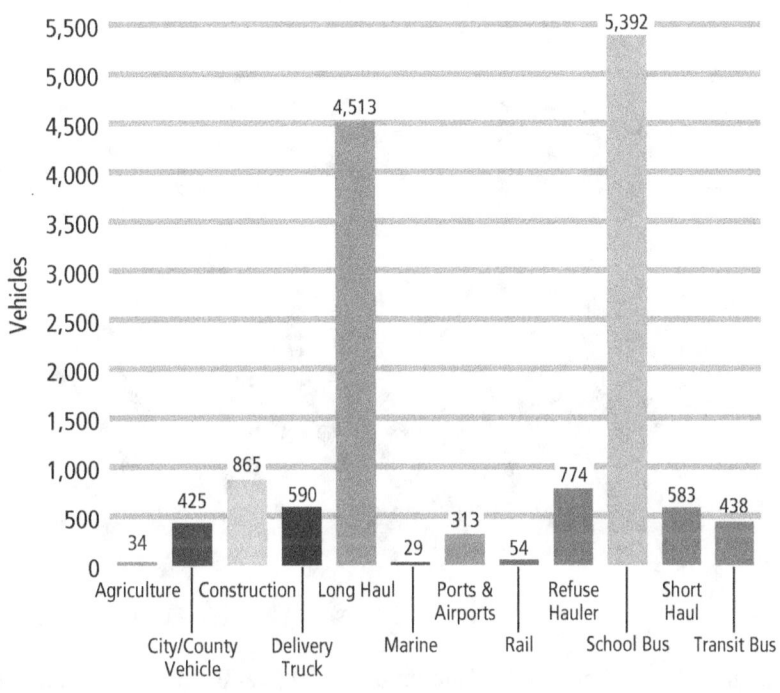

DERA Funding Supported a Wide Range of Technologies and Fuels. As shown in Figure 7, the FY 2008 projects represent a wide variety of verified technologies, cleaner fuels, and certified engine configurations, such as repowers, replacements, idle reduction technologies (e.g., auxiliary power units [APUs], shore power, truck stop electrification), biodiesel, alternative fuels, lower-sulfur fuels in nonroad equipment, and retrofit devices such as diesel oxidation catalysts and diesel particulate filters.

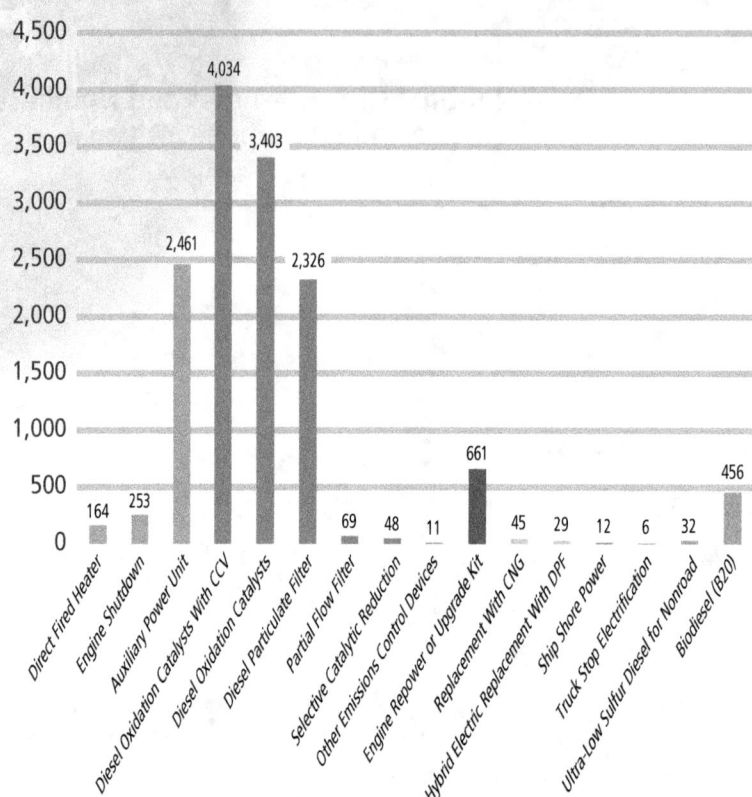

Figure 7. Technologies Employed in FY 2008 DERA Grants

DERA Funding Helped Reduce the Amount of Diesel Fuel Used. Table 3 shows the impressive amount of diesel fuel conserved by the idle reduction programs DERA funded. Taken together, the FY 2008 NCDC grants will save more than 3.2 million gallons of fuel per year, which corresponds to reducing 35,600 tons of CO_2 per year. This savings equates to more than $8 million a year in fuel savings for fleet operators (assuming $2.50 per gallon fuel). For comparison, freight trucks and locomotives consume 35 billion gallons of diesel fuel annually. Burning this fuel produces more than 350 million metric tons of carbon dioxide annually.

Table 3. Diesel Fuel Conserved Annually by FY 2008 DERA Projects With Idle Reduction Technologies

Program	CO_2 (annual tons)	Diesel Conserved (annual gallons)
National Clean Diesel Funding Assistance Program	8,200	736,000
National Clean Diesel Emerging Technologies Program	—	—
SmartWay Clean Diesel Finance Program	18,700	1,683,100
State Clean Diesel Grant and Loan Program	8,800	788,800
Total	35,600	3,207,900

Note: Values were rounded to the nearest 100 or to one significant figure. Sums may not add due to rounding. Calculation of CO_2 does not include CO_2 savings from engine repowers, replacements, or upgrades.

DERA Funding Results Are Relatively Cost-Effective. In addition to being successful in meeting their core goal, which is to reduce emissions, EPA's clean diesel programs are also cost-effective uses of federal funds, as shown in Table 4. As NCDC has developed over the years, clean diesel projects have become more cost-effective as the technology has developed, as more vendors are competing, and as stakeholders have become more experienced in implementing projects.

Table 4. Projected Cost-Effectiveness

Program	Projected NO_x	Projected $PM_{2.5}$	Projected HC	Projected CO	Projected CO_2 *	Program Federal Funding	Recipient Matching Amount
Federal costs per lifetime ton							
National Clean Diesel Funding Assistance Program	$2,000	$27,700	$15,300	$4,000	$300	$29,039,803	$34,749,839
SmartWay Clean Diesel Finance Program	$400	$9,000	$7,300	$2,100	$10	$3,390,000	$19,100,000**
State Clean Diesel Grant and Loan Program	$600	$16,700	$6,600	$2,300	$100	$12,994,080	$6,300,160
National Clean Diesel Emerging Technologies Program						$3,776,117	$1,130,927
Federal grant dollars per lifetime ton for three programs combined	$1,000	$20,600	$10,500	$3,100	$100		
Grant costs*** per lifetime ton							
Total dollars per lifetime ton*** for three programs combined	$1,600	$33,300	$16,900	$5,000	$200		

Note 1: Values were rounded to the nearest $100 or to one significant figure. Sums may not add due to rounding. Lifetime tons are not discounted.

* CO_2 was calculated for idle reduction technologies only and does not include reductions for repowers or replacements.

** Leveraged amounts; no matching was included in the grants, and this dollar amount is not included in the figure below.

*** Where cost includes federal funding, plus match, minus administrative costs assumed to be 15 percent, excluding leveraged amounts and the Emerging Technologies program.

Note 2: The Emerging Technologies program is designed to demonstrate new technologies. At this phase in the development of these technologies, costs would be expected to be higher than at a more fully mature commercial scale of deployment stage. Therefore, the costs comparisons with the other national program components shown in Table 4 have been omitted.

In a series of studies, EPA analyzed clean diesel strategies generally and evaluated the cost-effectiveness of using add-on retrofits on buses, trucks, and nonroad equipment using a standard set of assumptions.[9] Specifically, in these studies, EPA evaluated the cost-effectiveness of retrofitting school buses, freight trucks, and bulldozers with DOCs and catalyzed DPFs, two of the most common $PM_{2.5}$ emission reduction technologies for diesel engines.

---

[9] *The Cost-Effectiveness of Heavy-Duty Diesel Retrofits and Other Mobile Source Emission Reduction Projects and Programs* (EPA-420-B-07-006, May 2007)—Tables 3 and 4 provide comparisons of cost-effectiveness of other mobile source programs; and *Diesel Retrofit Technology: An Analysis of the Cost-Effectiveness of Reducing Particulate Matter and Nitrogen Oxides Emissions From Heavy-Duty Nonroad Diesel Engines Through Retrofits* (EPA-420-R-07-005, May 2007). The full reports are available on the EPA Web site: www.epa.gov/otaq/diesel/publications.htm.

- For nonroad equipment retrofit device cost-effectiveness:
  - PM: DOCs and DPFs for nonroad equipment ranged from $18,700 to $87,600 per ton of $PM_{2.5}$ reduced.
  - NO_x: SCR systems and engine upgrade kits were calculated to range from $1,900 to $19,000 per ton of NO_x reduced.

- For on-highway applications, such as school buses and trucks cost-effectiveness:
  - PM: DOCs ranged from $11,100 to $67,700 per ton of $PM_{2.5}$ reduced and DPFs retrofits ranged from $12,100 to $69,900 per ton of $PM_{2.5}$ reduced.
  - NO_x: Truck stop electrification ranged from $1,400 to $2,000 per ton of NO_x reduced, and APUs ranged from $2,700 to $3,500 per ton of NO_x reduced.

These estimates of cost-effectiveness for retrofit devices depend on a number of factors such as equipment activity, survival rates, effectiveness of the system, and costs.

For early retirements of vehicles and engine replacements or repowers, the new engines must meet EPA's more stringent regulations and thus are significantly cleaner. Cost-effectiveness estimates were made to support EPA's Clean Air Act regulatory programs.

- For nonroad vehicles:
  - PM: For EPA's Tier 1/Tier 2 standards, the cost-effectiveness is estimated at $2,000 per ton of $PM_{2.5}$ reduced.
  - NO_x:
    - For EPA's Nonroad Tier 4 standards, the cost-effectiveness is estimated at $1,010 per ton ($NO_x$ + nonmethane hydrocarbons [NMHC]).
    - For the locomotive and marine standards, the cost-effectiveness is $30 to $190 per ton ($NO_x$ + NMHC).

These PM-related original equipment manufacturer costs are comparable to or lower than those of installing retrofit technologies on an existing vehicle because of economies of scale in mass production and related factors. Depending on the age of the engine being retired, replacements and repowers can also improve fuel efficiency. Replacements can be cost-effective strategies for the oldest vehicles and vessels in the fleet, but there is often a significant financial barrier, because EPA grants only offset a portion of the overall cost of the vehicle. Several of the FY 2008 grants take advantage of the very cost-effective NO_x and PM reductions possible with locomotive and marine repowers.

Similarly, EPA anticipates that as emerging technologies become commercialized and more widely available, their cost-effectiveness too will be comparable to other verified technologies; however, at this early stage in their development, comparisons can be somewhat misleading, so they are not included in the analysis.

For comparison, for the PM NAAQS Regulatory Impact Assessment, for $PM_{2.5}$ reductions:

- The annualized cost for electrostatic precipitators for utility boilers was $3,000 to $15,000 per ton of $PM_{2.5}$ and $1,000 to $20,000 per ton for industrial boilers.

- The cost of fabric filters for these same units ranged from $2,000 to $100,000 per ton of $PM_{2.5}$.

- Area source controls' annualized costs are less than $100 per ton of $PM_{2.5}$.

When analyzing the effectiveness of the federal plus matching costs of the FY 2008 DERA projects, EPA found that they were within the expected range of cost-effectiveness. The average full program cost of the national, state, and SmartWay finance programs is about $33,300 per ton of $PM_{2.5}$ (including the matching amounts and assuming an upper limit on administrative costs of 15 percent, excluding leveraged amounts).

EPA analyzed the way in which federal dollars were leveraging other funding to achieve emission reductions. The reductions achieved over the lifetime of the FY 2008 DERA grants are being financed in different manners. In some cases, for instance, the federal share of the grant might pay for the entire cost of the retrofit device. In other cases, the federal share of the grant might provide a small amount, coupled with state incentive funds or private investment. In the SmartWay Clean Diesel Finance grants and in the other program grants that used the new loan provisions, relatively small federal investments on a per-unit basis attracted large private investment, especially where additional economic benefit in terms of fuel savings or the value of a newer vehicle was involved.

For the federal competitive national program:

- **National Clean Diesel Funding Assistance Program:** The average federal cost was estimated at $27,700 per ton of $PM_{2.5}$ reduced over the lifetime of the project and just under $2,000 per ton of NO_x reduced over the project lifetime.

- **SmartWay Clean Diesel Finance Program:** The average federal cost was estimated at $9,000 per ton of $PM_{2.5}$ reduced over the lifetime of the project and $400 per ton of NO_x reduced over the project lifetime. Federal funds leveraged significant private sector funding and fuel savings in this new approach.

For the State Clean Diesel Grant and Loan Program:

- The average federal cost was estimated at $16,700 per ton of $PM_{2.5}$ and $600 per ton of NO_x over the lifetime of the project. State program costs are lower than national program costs because many states have established programs that are generally more limited to specific lower-cost vehicle categories or have leveraged amounts not captured here.

For the three programs together:

- The grants will reduce approximately 46,000 tons of NO_x and 2,240 tons of $PM_{2.5}$ over the programs' lifetimes in addition to substantial reductions in HC and other pollutants.

- The weighted average federal cost is about $25,000 per ton of $PM_{2.5}$ and $1,500 per ton of NO_x reduced over the lifetime of the program.

DERA Funding Is Yielding Tangible Public Health Benefits. The DERA programs reduce emissions that affect people's everyday lives, especially for sensitive populations such as children, the elderly, those with heart and lung disease, outdoor workers, and people who exercise outdoors. The diesel emission reductions resulting from the FY 2008 grants for NO_x and PM will total approximately 46,000 and 2,200 tons, respectively, by 2031, yielding a variety of health benefits. Fewer emissions leads to less air pollution, resulting in fewer respiratory symptoms, reduction in the use of medication for asthma, and fewer visits to emergency departments or health care providers. Parents miss fewer work days caring for their children or their own respiratory symptoms.

- The health benefits from the FY 2008 grant projects will range from a net present value of $580 million to $1.4 billion (2006 dollars), assuming a 3-percent discount rate throughout the lifetime of the program.

- These benefits include an estimated 95 to 240 avoided premature deaths.[10]

Ozone-related health benefits, cancer risk, and other welfare effects were not quantified in this analysis.

---

[10] The estimates are based on the peer-reviewed studies on the relationship between $PM_{2.5}$ and premature mortality from:

Pope, C.A., III, R.T. Burnett, M.J. Thun, E.E. Calle, D. Krewski, K. Ito, and G.D. Thurston. 2002. Lung Cancer, Cardiopulmonary Mortality, and Long-Term Exposure to Fine Particulate Air Pollution. *Journal of the American Medical Association*. 287: 1132-1141.

Laden, F., J. Schwartz, F.E. Speizer, and D.W. Dockery. 2006. Reduction in Fine Particulate Air Pollution and Mortality. *American Journal of Respiratory and Critical Care Medicine*. 173: 667-672.

For more information about this approach, please see EPA's 2008 *Technical Support Document: Calculating Benefit Per-Ton Estimates*, Ozone NAAQS Docket #EPA-HQ-OAR-2007-0225-0284. The premature mortality estimates are generated using $PM_{2.5}$ co-benefits data from EPA's 2006 *Ozone Final Regulatory Impact Analysis (RIA)*, NAAQS for Particulate Matter, Chapter 5 Benefit Analysis and Results. These techniques are described in a memo to the file: Memo From Michael Wolfe, Ken Davidson, Rosalva Tapia, Kuang Wei, and Jen Went to Jim Blubaugh, Manager, Innovative Strategies Group, EPA Office of Transportation and Air Quality. March 25, 2009. Energy Policy Act of 2005 Diesel Emissions Reduction (DERA) Program FY 2008: Benefit and Cost-Effectiveness Calculation Methodologies.

4. Resources That Support the National Clean Diesel Campaign

Partners Fulfill Commitments and Measure Progress

WHILE reducing diesel emissions is the first priority, producing technical and programmatic support materials for stakeholders is a necessity as well. To ensure the successful implementation of NCDC, EPA developed the Diesel Emissions Quantifier (DEQ) calculator; guidance documents; public outreach materials including a dynamic Web site, an idle reduction campaign, and a Scholastic *Magic School Bus* book; and a Web-based tool kit for state and local programs.

DIESEL EMISSIONS QUANTIFIER

To help state and local governments, fleet owners, school districts, municipalities, contractors, port authorities, and others in estimating emission reductions and cost-effectiveness for clean diesel projects, EPA created an interactive, Web-based calculator tool called the DEQ. Users enter specific information about a fleet, and emission and cost-effectiveness calculations are made based on existing EPA models and guidance. The DEQ performs real-time calculations and estimates project-level emissions for $PM_{2.5}$, NO_X, HC, CO, and CO_2. To support users of this tool, EPA also created a user guide and held interactive training webinars.

Ultimately, the tool became a critical component of NCDC, as most grant applicants rely on it to estimate their emission reductions. Grant applicants are required to evaluate the quantifiable and unquantifiable benefits of the emission reductions of their proposed projects in each grant application, using a methodology approved by EPA or the National Academy of Sciences.

AGENCY GUIDANCE FOR CALCULATING COST-EFFECTIVENESS

EPA published two peer-reviewed technical papers analyzing the cost-effectiveness of reducing emissions from both on-highway and nonroad diesel engines. (The full reports are available at: www.epa.gov/otaq/diesel/publications.htm.)

• *The Cost-Effectiveness of Heavy-Duty Diesel Retrofits and Other Mobile Source Emission Reduction Projects and Programs* (EPA-420-B-07-006, May 2007).

- *Diesel Retrofit Technology: An Analysis of the Cost-Effectiveness of Reducing Particulate Matter and Nitrogen Oxides Emissions From Heavy-Duty Nonroad Diesel Engines Through Retrofits* (EPA-420-R-07-005, May 2007).

In these papers, EPA evaluated the costs and emission benefits of retrofitting school buses, freight trucks, and bulldozers with DOCs and catalyzed DPFs, two of the most common PM emission reduction technologies for diesel engines. The findings from these studies indicate that retrofits can be a cost-effective way to reduce air pollution compared with other EPA programs.

GUIDANCE ON STATE IMPLEMENTATION PLANS

EPA published guidance to help state and local air quality and transportation planners satisfy three EPAct 2005 Section 795 provisions: 1) quantify the emission reductions from retrofitting diesel vehicles, engines, and equipment; 2) appropriately include the emission reductions from diesel retrofits in state implementation plans (SIPs) to help demonstrate progress toward, attainment of, or maintenance of NAAQS; and 3) appropriately include emission reductions from diesel retrofits in transportation conformity or general conformity.

- *Diesel Retrofits: Quantifying and Using Their Benefits in SIPs and Conformity—Guidance for State and Local Air and Transportation Agencies* (EPA-420-B-06-005, June 2006).
- *Truck Guidance: Guidance for Quantifying and Using Long-Duration Truck Idling Emission Reductions in State Implementation Plans and Transportation Conformity* (EPA-420-B-04-001, January 2004).
- *Locomotive Guidance: Guidance for Quantifying and Using Long-Duration Switch Yard Locomotive Idling Emission Reductions in State Implementation Plans* (EPA-420-B-04-002, January 2004).

PUBLIC OUTREACH MATERIALS

EPA has created a robust national Web site (www.epa.gov/cleandiesel) and myriad public outreach materials to assist partners and others with diesel emission reduction projects. In addition, each Regional Collaborative maintains a Web site with regional priorities and resources.

As a sign of the national Web site's utility, more than 80,000 users visited the site in FY 2008. The site is continually updated with vital new information and materials, tailored to meet stakeholder needs. Its design includes a searchable list of demonstration projects, case studies, technology information, listservs, a trucker portal (www.epa.gov/otaq/diesel/truckers.htm), sector-specific content (www.epa.gov/cleandiesel/ports), and other helpful links, including links to Regional Collaborative Web sites.

The breadth of EPA's reach in communicating information about clean diesel and NCDC to the public in FY 2008 can be measured by the following statistics:

- 100 diesel conferences supported by EPA via financial sponsorships, promotion, and participation, including the delivery of 12 major speeches by EPA's senior executives.

- 19 press releases issued highlighting clean diesel accomplishments and their benefit to the public.

- 40,125 program brochures distributed to stakeholders.

- 650 Clean School Bus USA idle reduction training videos in use.

- 300 Clean Construction USA videos viewed.

- 128,000 copies of Scholastic's *The Magic School Bus Gets Cleaned Up* books distributed.

NCDC JOINS FORCES WITH INTERNATIONAL PARTNERS

Using funding from non-DERA sources, partnering with local governments and nongovernmental organizations, and supported by U.S. technology vendors, EPA led demonstration projects in Mexico City, Mexico; Bangkok, Thailand; Santiago, Chile; Beijing, People's Republic of China; and Pune, India.

EPA retrofitted about 20 vehicles in each city to demonstrate the emission reductions obtainable with various devices. The projects helped educate government and nongovernment staff about clean diesel technologies and aim to build the capacity so they can conduct retrofits on their own. Following the Beijing demonstration project, the city of Beijing independently retrofitted more than 5,000 diesel vehicles in preparation for the 2008 Summer Olympics. EPA also worked with the U.S. Department of Transportation's Maritime Administration on a project with the Port of Shanghai that resulted in a broader Pacific Ports Initiative, promoting cleaner diesel technologies among key trading partners and their ports.

Also on the international front, EPA launched a cooperative project among the Port of Rotterdam, the Port of New York and New Jersey, a terminal operator with facilities in both ports, and technology providers to develop hydraulic hybrid cargo handling equipment. This technology promises to dramatically improve fuel economy while reducing PM and NO_x.

Lastly, EPA hosted a workshop for Canadian federal and provincial staff in June 2008. The workshop transferred knowledge and built relationships with appropriate Canadian government personnel so that they might learn from the NCDC efforts and be better equipped to launch their own program.

These international actions are consistent with DERA Section 795(d), "Outreach and Incentives; International Markets."

Tool Kit for State and Local Government Partners

To support state, regional, and local governments in improving air quality and public health through diesel engine emission reduction efforts, EPA developed a Web-based tool kit that compiles examples of program components they could model (www.epa.gov/otaq/diesel/slt/basicinfo.htm). The content focuses on designing programs, communicating success, identifying funding, and evaluating program results. This tool kit is just one example of how NCDC operates as a shared responsibility between EPA and state and local air quality agencies.

5. Remaining Challenges
Adding Verified Technologies and Incentives Could Enhance Implementation

EPA's advance work in developing guidance and materials, offering educational tools to partners developing clean diesel programs, and communicating consistently and clearly with partners and stakeholders was essential to the success of the first year. At the same time, the Agency's ability to troubleshoot problems that arose and working within the collaborative and sector networks to improve and adjust the program along the way ensured that it met its mission.

The challenges encountered in implementing the first year of this program were relatively few. At the end of the year, however, two challenges remain:

1. The number of verified technologies for nonroad and marine engines is too few.

While the list of technologies available on EPA and CARB's lists for on-highway applications is long, the technologies available for nonroad and marine applications are limited. The agencies have identified a lack of laboratory facilities and field testing instruments capable of evaluating marine technologies as the main barrier. Both agencies have been focusing their efforts on increasing the availability of nonroad retrofit technologies, closely working with technology manufacturers to encourage additional technology verification for nonroad and marine equipment.

There are also some important gaps in diesel retrofit technologies for older heavy-duty diesel trucks. While EPA and CARB have verified numerous heavy-duty diesel truck retrofit technologies, the low-cost devices only reduce PM emissions by a limited amount, and the more effective devices might not work as well with the older trucks' duty cycles and business models. In some cases, the value of the retrofit device can exceed the remaining value of an old truck. Accordingly, EPA will continue to work with diesel retrofit manufacturers to identify cost-effective solutions for older diesel trucks.

Responding to this need, EPA held a Retrofit Technology Verification Workshop in Washington, D.C., on December 13, 2007. The workshop was an overwhelming success, with nearly 100 people attending, including technology vendors, original equipment manufacturers, fuel and fuel additive makers, diesel engine technology organizations, and state and local governments such as CARB.

This workshop spurred additional verification activity among manufacturers. In 2008, EPA approved seven additional technologies for nonroad applications and 13 additional technologies for on-highway applications. In addition, EPA approved six emerging technologies (three nonroad and three on-highway applications), which will continue in the verification process and could well be verified within the next two years.

Despite these gains, EPA had to remove a number of technologies from its Technology Verification List in December 2008, due to the implementation of a new, more stringent limit on the amount of unwanted nitrogen dioxide (NO_2) that retrofit devices are allowed to produce. EPA anticipates that some of the removed technologies will soon meet the new requirement and the Agency will be able to reinstate them.

2. New incentives are needed to retire the oldest and dirtiest engines.

As EPA's Clean Diesel regulations under the Clean Air Act phase in, new engines that must meet significantly more stringent standards are increasingly available in the marketplace. The challenge is to create additional incentives for private fleet owners to replace their older diesel engines with new engines that meet the new standards.

Under the current law, national program grants cannot go toward private fleet owners or to projects for which there is a federal, state, or local mandate. First, it is a challenge to encourage private fleets to apply, as they must partner with eligible entities, creating a barrier to participation especially for small businesses that may own or operate the oldest and dirtiest fleets. Second, routine attrition of old vehicles is considered to be compliance activity with Clean Air Act requirements and currently is not eligible for funding. Accordingly, early replacement projects can receive federal grants, but not projects that might target the oldest, dirtiest engines that are already past their useful life and should have been replaced through normal attrition but are nevertheless still in use. These dual challenges are especially great for small business owner/operators for equipment such as drayage trucks that serve seaports, tug and tow vessels, and construction equipment.

EPA will continue to work to identify incentives utilizing the innovative finance and loan provisions of the statute and our SmartWay Transport Partnership, as well as to fashion additional incentives together with our state and local government partners.

6. Lessons Learned and Needs for the Future
Continue, Improve, and Expand

HILE the DERA program has achieved significant emission reductions from diesel engines in its first year, a tremendous amount of work remains, and substantial investments are needed.

LESSONS LEARNED

Lessons learned from the first year include the following:

- **EPA and others must continue efforts to utilize existing national and regional networks to further improve the program results.** In order to expedite the program, EPA developed materials to communicate the program options, technologies, goals, and schedules. Providing clear information and catalyzing networks allowed all parties to engage in the process and move projects forward. As the programs grow, EPA must promote efforts that allow agencies to share experiences to encourage action. To this end, EPA will work to strengthen existing regional diesel collaboratives.

- **EPA must continue to cultivate state and local leadership.** All 50 states elected to participate in the first year of the program. For some, this was just the beginning of their work in this program area. To ensure future success, EPA must continue to work with state agencies and the District of Columbia to build comprehensive clean diesel programs. The results of the program can be seen at the state and local levels where leadership from state and local agencies, nonprofit organizations, and private organizations engages interested fleets in the program and works with them to achieve results. EPA must encourage more organizations to take this role as both an environmental initiative and an economic opportunity.

- **EPA and other organizations must continue to involve the private sector.** The private sector plays a key role in implementing the program. Organizations that supply eligible technologies must continue to work to improve their technologies and identify opportunities for fleets that are willing to take action. In addition, EPA and other agencies should encourage private fleets in priority areas to take proper steps to address the air pollution from their fleets. The SmartWay Transport Partnership, for example, has created a program for private fleets in the freight sector. In exchange for making a commitment to reduce emissions and fuel usage, EPA provides technical advice, public recognition, and the opportunity to become preferred carriers for SmartWay shippers. Through corporate leadership to reduce emissions and reduce fuel usage, the DERA program has a great opportunity for expansion. In order to accomplish this, corporate champions must step forward and make measurable commitments.

- **Organizations must continue to develop and advance verified nonroad and marine technologies.** Opportunities continue to arise in new program areas as the program grows. Vendors must recognize these opportunities and work to create improved options to address the multitude of emission sources covered under the legislation, especially in the nonroad and marine sectors. EPA will expand the emerging technology and verification programs to promote innovative technologies that reduce priority pollutants and conserve fuel.

While the verified technologies list available on EPA's and CARB's Web sites for on-highway applications is long, there are limited products verified for nonroad, marine, or older diesel trucks. There are limited products available to reduce NO_x for any category of engine. Adequate laboratory facilities and field testing instruments capable of evaluating PM emission reductions from marine technologies will be needed. EPA has been working closely with manufacturers to verify technologies and fill these gaps. The nonroad market is complicated by the number and diversity of nonroad equipment types, by the range of horsepowers and engine types involved, and by the varying usage and duty cycles of the equipment. There are many opportunities, however, to provide cost-effective diesel emission reduction technologies to meet local air quality needs.

NEXT STEPS

Moving forward in the program, EPA commits to the following:

- Continue to work aggressively to reduce pollution from diesel engines across the country by partnering with key stakeholders to promote clean diesel strategies.
- Target current PM and ozone nonattainment areas where clean diesel strategies will have the greatest public health impact.
- Provide assistance to state and local governments in developing their own clean diesel programs.
- Continue to provide high-quality data to states that depend on the performance of diesel emission reduction strategies in their air quality plans, through in-use testing.
- Continue to work cooperatively with CARB and other states to provide a robust list of clean diesel technology options for partners.
- Continue to confirm the emission performance of verified technologies in the field.
- Continue to develop innovative financing approaches for stretching federal dollars to achieve maximum diesel emission reductions, especially where fuel savings can create a business case for low-cost loans or other financial incentives.
- Develop timely educational materials to continue to build awareness of clean diesel opportunities.

In summary, the first year of implementation for the Diesel Emissions Reduction Program has been extremely successful. By harnessing broad support from environmental organizations, state and local governments, industry, technology vendors, and other groups, EPA has been able to magnify the available dollars and strategically fund important emission reduction and fuel saving projects.

In total, EPA awarded 119 grants in the first year of the DERA program. More than 14,000 vehicles and pieces of nonroad equipment in a wide array of sectors will be cleaner as a direct result of this program. New clean diesel programs have been established in every state and technologies have been advanced for the future. The implementation of the EPAct 2005's Diesel Emissions Reduction Program is a critical step toward reaching our national clean air goals and protecting public health and the environment for all Americans for generations to come.

APPENDIX A: National Clean Diesel Funding Assistance Program, FY 2008 Grants

State	Grant Recipient	EPA Award	Match	Sector(s)	Technology Types*
AL	Alabama State Port Authority	$750,000	$1,580,000	Ports, Rail	Replacement
AZ	City of Phoenix	$553,629	$732,000	Utility	Engine Replacement, DPF
CA	CALSTART, Inc.	$895,827	$1,015,325	Construction, Utility, Transit Bus	CNG, DPF, B20 Fuel
CA	Kern County Superintendent of Schools	$540,000	$540,000	School Bus	CNG Replacement
CA	Sacramento Metropolitan Air Quality Management District	$553,630	$2,581,600	Utility, School Bus	DPF, Engine Replacement
CA	South Coast Air Quality Management District	$1,000,000	$7,095,900	Long-Haul Trucks	DPF
CO	City and County of Denver	$200,000	$87,059	Utility, Construction	Heater
CO	Colorado Department of Public Health	$400,000	—	School Bus	DOC, Closed Crankcase Ventilation (CCV) System and Engine Preheater
CO	Regional Air Quality Council	$455,645	—	School Bus	DOC, CCV System
CT	Connecticut Department of Environmental Protection	$50,000	—	Utility, Construction	Diesel Oxidation Catalyst, DPF, Ultra-Low Sulfur Diesel
DC	Metropolitan Washington Council of Governments	$598,516	$193,983	Construction	DPF, Repower, Engine Upgrade
FL	City of St. Petersburg	$666,510	$5,745	Utility, Construction	DPF, B20 Fuel
GA	Georgia Ports Authority	$250,000	$33,075	Ports	DOC, CCV System
ID	Idaho Department of Environmental Quality	$500,000	$500,000	School Bus	DPF, Partial Flow Filter, DOC
IL	Chicago Public Schools	$373,909	$92,000	School Bus	Retrofit (DOC, DPF)
IL	Illinois EPA	$678,604	$654,699	School Bus, Transit Bus, Utility	Retrofit (DOC, DPF, CCV System), Auxiliary Power Unit, Direct-Fired Heater, Hybrid Replacement

* Refer to page 56 for a list of acronyms and abbreviations.

Continued on page 44

Appendix A. National Clean Diesel Funding Assistance Program, FY 2008 Grants *(continued)*

State	Grant Recipient	EPA Award	Match	Sector(s)	Technology Types
IN	Indiana Department of Environmental Management	$334,500	$61,257	Transit Bus	Retrofit (DOC, DPF), APU
IN	Northwest Indiana Forum Foundation Inc.	$164,032	$173,432	Construction	Repower
KS	Kansas Department of Health and Environment	$1,525,524	$1,871,500	Trucks and Transit Buses	DOC w/ CCV System
KY	Kentucky Clean Fuels Coalition	$473,939	$2,160,000	Ports	Vehicle Replacement
MA	Massachusetts Port Authority (Massport)	$400,000	$147,815	Ports	Shore Power
MA	NESCAUM	$400,000	$100,000	Construction	DPF
MA	NESCAUM	$535,250	—	Rail	Auxiliary Power Unit
MD	Maryland Environmental Services	$361,951	$25,909	Ports	DPF
MD	Montgomery County Public Schools	$699,501	$7,350	School Bus	DPF
MI	Lenawee Intermediate School District	$154,381	—	School Bus	DPF
MI	Michigan Clean Energy Coalition	$250,000	$250,580	Construction	Repower
MI	NextEnergy Center	$250,000	$650,500	Long Haul Trucks	APU
MI	Laker School District	$251,100	$253,220	School Bus	Bus Replacement; Idle Reduction Technology, B20 Fuel, Heater
MN	Minnesota Environmental Initiative	$400,000	—	School Bus, Utility	Retrofit (DOC, CCV System)
MO	Grace Hill Settlement House	$454,849	—	School Bus	Crankcase Filtration System (CFS)
MO	Missouri Department of Natural Resources	$726,227	$621,550	Construction	DOC, CCV System, APU, Engine Shutdown
NC	North Carolina Department of Environment and Natural Resources	$750,000	—	Construction	Repower, Replacement
NH	Manchester Transit Authority	$229,703	$5,652	School Bus, Construction, Utility	DOC, CCV System, S-Bar Heater, B20 Fuel
NJ	New Jersey Motor Truck Association	$503,285	—	Long-Haul Trucks	APU, Bunk Heater, DOC, DPF
NY	Board of Cooperative Educational Services of Ulster County	$130,690	—	School Bus	Vehicle Replacement
NY	Capital District Transportation Authority	$125,000	—	Transit Bus	DPF
NY	Environmental Defense Fund	$400,000	$2,025,000	Utility	Vehicle Replacement (Hybrid-Electric)

Appendix A. National Clean Diesel Funding Assistance Program, FY 2008 Grants *(continued)*

State	Grant Recipient	EPA Award	Match	Sector(s)	Technology Types
NY	Erie County Department of Environmental Planning	$523,395	$72,278	School Bus	DOC, CCV System, Diesel-Fired Engine Warm-Up Heater
NY	Middle Country Central School District	$359,305	$361,305	School Bus	Vehicle Replacement (CNG)
NY	Scarsdale Union Free School District	$346,240	$29,020	School Bus	DOC, CCV System
NY-NJ	Port Authority of New York and New Jersey (PANYNJ)	$280,500	$80,500	Utility	DPF, Flow Through Filter, DOC
NY-NJ	PANYNJ	$750,000	$1,250,000	Ports	Flow Through Filter, DPF
OH	Ohio Environmental Council	$394,589	$29,571	Ports, Utility, Refuse Haulers	Retrofit (DOC, DPF, CCV System); APU
OH	Ohio Indiana Clean Diesel Collaboration	$412,554	$81,344	Utility	DPF
OH	Stark County Educational Service Center	$465,364	$89,905	School Bus	Retrofit (DPF, CCV System, DOC), Replacement
OR	City of Portland	$498,726	$58,658	Ports, Construction	DOC, Exhaust Retrofit, Ultra-Low Sulfur Diesel
PA	City of Philadelphia	$750,000	$2,250,000	Utility	CNG Vehicle Replacement
PA	Pennsylvania Department of Transportation	$219,434	$24,381	Rail	APU
SC	South Carolina Port of Charleston	$735,002	$963,502	Utility, Construction	Engine Repower, Ultra-Low Sulfur Diesel
SD	Sioux Falls School District	$300,000	$300,000	School Bus	Replacement
TN	Knox County Government	$163,871	$23,056	Utility, Construction, Agricultural	Emission Control, Ultra-Low Sulfur Diesel
TX	North Central Texas Council of Governments	$750,000	$750,000	Trucks	APU
TX	North Central Texas Council of Governments	$750,000	$1,500,000	Construction	Replacement
TX	Texas Commission on Environmental Quality	$500,000	—	School Bus	DOC, DPF, Partial Flow Filter, CCV System
UT	Utah Department of Air Quality	$400,000	$4,000	School Bus	DOC and CCV System
VA	Port of Norfolk	$647,457	$1,732,600	Ports and Rail	Engine Repower
VT	Chittenden Solid Waste District	$205,523	$616,569	Utility	Vehicle Replacement

Appendix A. National Clean Diesel Funding Assistance Program, FY 2008 Grants *(continued)*

State	Grant Recipient	EPA Award	Match	Sector(s)	Technology Types
WA	Puget Sound Clean Air Agency	$850,000	$318,000	Ports	Vehicle Replacement, DPF, Partial-Flow Filter, Crankcase Ventilation Filter, DOC
WI	Wisconsin Department of Transportation	$750,000	$750,000	Construction	Repower
Total		**$29,038,162**	**$34,749,839**		

APPENDIX B: National Clean Diesel Emerging Technologies Program, FY 2008 Grants

State	Grant Recipient	EPA Award	Match	Sector(s)	Vehicle Types	Technology Types
CA	South Coast Metropolitan Air Quality Management District	$900,000	$858,614	Delivery Truck	Class 8a (33,001–60,000 lbs)	Selective Catalytic Reduction (SCR)
MD	Montgomery County Maryland	$400,000	$88,168	Delivery Truck	Class 6 (19,501–26,000 lbs)	SCR
TX	Center for Transportation and the Environment	$300,000	—	Short Haul	Class 8a (33,001–60,000 lbs)	Other Emissions Control Devices
TX	Texas Transportation Institute	$500,000	$64,200	Construction	Graders	SCR
TX	University of Houston	$500,000	$119,945	Construction	Tractors, Loaders, Backhoes	SCR
WA	Puget Sound Clean Air Agency	$700,000	—	Marine	Marine	Engine Upgrade
Total		$3,300,000	$1,130,927			

For the FY 2008 National Clean Diesel Emerging Technology Program competition, EPA approved six technologies, as shown below. These included two technologies for marine vessels and SCR technologies for on-highway vehicles and nonroad equipment that expand technological options to reduce NO_x by approximately 65 percent. This list of EPA's emerging technologies is updated as information is submitted from manufacturers, and EPA anticipates that the types and numbers of technologies will continue to grow.

Manufacturer	Technology	Application	Reductions (%)			
			PM	CO	NO_x	HC
Caterpillar, Inc.	Engine Upgrade Kit	Marine	25	N/A	N/A	N/A
ESW Canada	DOC, CCV System	Marine	25	70	N/A	25
Johnson Matthey	Urea-Based SCR System, DPF	On-Highway	90	85	65	95
Nett Technologies, Inc.	Urea-Based SCR System	Nonroad	20	60	65	60
Tinnerman/Shadowood	Reformer, LNT, SCR, DPF	On-Highway	90	90	65	90
Truck Emission Control Technologies Inc.	DOC, DPF, EGR	On-Highway	50	70	40	60

APPENDIX C: SmartWay Clean Diesel Finance Program, FY 2008 Grants

Grant Recipient	Project Overview
Cascade Sierra Solutions Everybody Wins USA Lease Program Grant Amount: $1.13 million	**Loan Interest Rate:** 8% to 11%, includes 7% for financing cost, 1% collection fee, and 1% to 3% reservation loss. **Loan Payback Period:** Three years. **Eligible Activities:** Installation and leasing of EPA SmartWay or CARB-verified emission and idle reduction technologies. The truck owners will purchase the equipment for $10 at the end of the loan period. **Program Goal:** Establish a leasing program for the installation of emission reduction technology and idle reduction technology on 1,700 trucks.
Community Development Transportation Lending Services, Inc. National Low Interest Revolving Loan Fund Grant Amount: $1.13 million	**Interest Rates:** 5.5% to 8.5%, varies based on, among other things, each applicant's business history, available collateral, cash flow, and credit score. **Loan Payback Period:** Three to six years, varies based on, among other things, business history, available collateral, and cash flow. **Eligible Activities:** Purchase used trucks to upgrade with EPA SmartWay or CARB-verified idle reduction and/or emission reduction technology. Purchase used trucks that are upgraded with EPA SmartWay or CARB-verified idle reduction and/or emission reduction technology. **Program Goal:** Establish a revolving loan fund for the purchase of used trucks that have SmartWay upgrades or will receive SmartWay upgrades. **Service Fee:** 2% to 3%.
Owner-Operator Independent Drivers Association Foundation Innovative Financing Grant Amount: $1.13 million	**Loan Interest Rate:** Varies based on market Solid Waste Alternative Program rate. **Eligible Activities:** Purchase and installation of emission reduction and idle reduction technologies verified by EPA and/or CARB. **Program Goal:** Offer rebates to selected applicants/truck owners to install emission reduction and/or idle reduction equipment on their trucks. The first rebate to the truck owner will be for the cost of the equipment installation. Then, every six months, depending on idle/fuel use reductions demonstrated, the truck owner will receive additional rebates until 40% of the total cost of the equipment is paid.

APPENDIX D: State Clean Diesel Program, FY 2008 Grants

State	Grant Recipient	Total Federal Funding Allotment	Sector(s)	Technology Types
AK	Alaska Department of Environmental Conservation	$295,320	Construction	Research, Retrofit
AL	Alabama Department of Environmental Management	$196,880	Rail	Repower
AR	Arkansas Department of Environmental Quality	$295,320	School Bus, Long Haul Trucks, Construction	Retrofitted, Idle Reduction
AZ	Arizona Department of Environmental Quality	$196,880	Short Haul	Truck Stop Electrification
CA	California Air Resources Board	$295,320	Long Haul Trucks	Retrofit
CO	Colorado Department of Public Health and Environment	$196,880	School Bus	Retrofit
CT	Connecticut Department of Environmental Protection	$295,320	School Bus, Long Haul Trucks	Idle Reduction
DE	Delaware Department of Natural Resources and Environmental Control	$295,320	Refuse Hauler	Retrofit
FL	Florida Department of Environmental Protection	$295,320	Long Haul Trucks, School Bus	Retrofit, Idle Reduction
GA	Georgia Department of Natural Resources	$295,320	School Bus, Construction	Retrofit
HI	Environmental Health Administration	$196,880	School Bus, Utility	Retrofit
IA	Iowa Department of Natural Resources	$196,880	School Bus	Replacement
ID	Idaho Department of Environmental Quality	$196,880	School Bus	Retrofit
IL	Illinois Environmental Protection Agency	$295,320	Long Haul Trucks	Idle Reduction
IN	Indiana Department of Environmental Management	$295,320	Long Haul Trucks	Retrofit
KS	Kansas Department of Health and Environment	$295,320	Long Haul Trucks	Replacement, Idle Reduction

Appendix D. State Clean Diesel Program, FY 2008 Grants *(continued)*

State	Grant Recipient	Total Federal Funding Allotment	Sector(s)	Technology Types
KY	Kentucky Division for Air Quality	$196,880	School Bus	Retrofit
LA	Louisiana Department of Environmental Quality	$295,320	Marine	Repower
MA	Massachusetts Department of Environmental Protection	$295,320	Refuse Hauler, Rail, Utility	Retrofit, Replacement
MD	Maryland Department of the Environment	$295,320	Transit Bus	Retrofit
ME	Maine Department of Environmental Protection	$295,320	School Bus, Long Haul Trucks	Idle Reduction
MI	Michigan Department of Environmental Quality	$295,320	School Bus	Retrofit
MN	Minnesota Pollution Control Agency	$295,320	School Bus, Long Haul Trucks	Retrofit, Idle Reduction
MO	Missouri Department of Natural Resources	$295,320	Long Haul Trucks, Utility	Replacement, Idle Reduction
MS	Mississippi Department of Environmental Quality	$295,320	Utility Vehicle, School Bus	Retrofit
MT	Montana Department of Environmental Quality	$295,320	School Bus, Transit Bus	Retrofit, Repower
NC	North Carolina Department of Environment and Natural Resources	$295,320	Transit Bus, School Bus, Rail, Construction, Refuse Haulers	Retrofit, Idle Reduction, Repower
ND	North Dakota Department of Health	$196,880	School Bus	Replacement
NE	Nebraska Department of Environmental Quality	$196,880	School Bus, Refuse Hauler, City/County Vehicle, Transit Bus, Construction	Retrofit
NH	New Hampshire Department of Environmental Science	$196,880	Transit Bus, Construction, Utility	Retrofit
NJ	New Jersey Department of Environmental Protection	$295,320	Construction, School Bus	Retrofit
NM	New Mexico Environment Department	$196,880	School Bus	Retrofit
NV	Nevada Division of Environmental Protection	$295,320	School Bus	Retrofit
NY	New York State Department of Environmental Conservation	$295,320	Marine	Replacement
OH	Ohio Environmental Protection Agency	$295,320	School Bus	Retrofit
OK	Oklahoma Department of Environmental Quality	$295,320	School Bus	Retrofit

Appendix D. State Clean Diesel Program, FY 2008 Grants *(continued)*

State	Grant Recipient	Total Federal Funding Allotment	Sector(s)	Technology Types
OR	Oregon Department of Environmental Quality	$295,320	School Bus	Retrofit
PA	Pennsylvania Department of Environmental Protection	$295,320	Ports, Airports	Repower
RI	Rhode Island Department of Environmental Management	$196,880	Ports, Airports	Retrofit
SC	South Carolina Department of Health and Environmental Control	$295,320	School Bus, Construction	Retrofit, Repower
SD	South Dakota Department of Environment and Natural Resources	$196,880	School Bus	Retrofit
TN	Tennessee Department of Environment and Conservation	$196,880	Long Haul Trucks	Idle Reduction
TX	Texas Commission on Environmental Quality	$295,320	School Bus	Retrofit
UT	Utah Department of Environmental Quality	$295,320	School Bus	Retrofit
VA	Virginia Department of Environmental Quality	$196,880	Utility Vehicle	Retrofit
VT	Vermont Department of Environmental Conservation	$196,880	School Bus	Retrofit, Idle Reduction
WA	Washington State Department of Ecology	$295,320	School Bus	Retrofit
WI	Wisconsin Department of Natural Resources	$295,320	Refuse Hauler, Construction	Retrofit, Repower
WV	West Virginia Department of Environmental Protection	$196,880	Transit Bus	Replacement
WY	Wyoming Department of Environmental Quality	$196,880	School Bus	Repower
Total		**$12,994,080**	**Match**	**$6,300,160**

APPENDIX E: National Program Evaluation Criteria

For the National Clean Diesel Funding Assistance Program, all grant competitions used criteria and points similar to those summarized in the table below, consistent with the priorities described in Section 792 of the Energy Policy Act, Subtitle G (see page 20), and with Agency policy.

National Clean Diesel Funding Assistance Program

For the National Clean Diesel Funding Assistance Program, the seven EPA Regional Collaboratives described in Appendix F held grant competitions, following Agency policy. Regions evaluated each Regional RFP following the same general criteria, supplementing with regional priorities worth 10 points in the final criterion.

EPA devoted the majority of the funding available to the National Clean Diesel Funding Assistance Program and allocated it to its 10 regional offices to award as competitive grants in support of clean diesel projects. EPA distributed funds across the regions based on a funding formula that included a base amount and considered population, diesel emission inventories, and other factors consistent with the objectives listed in the statute Section 792.

EPA's regions addressed certain regional priorities in their competitions, while following a model RFP to ensure that all of EPA's actions adhered to the statutory requirements and EPA policy.

National Clean Diesel Emerging Technologies Program

For the National Clean Diesel Emerging Technologies RFP, EPA employed similar criteria to those in the table below; however, EPA required that the projects propose to use technologies on EPA's Emerging Technology List by the closing date of the RFP. Criterion 9 was not applicable.

SmartWay Clean Diesel Finance Program

For the SmartWay Clean Diesel Finance RFP, EPA used criteria similar to those in the table below; however, Criterion 9 was not applicable.

Appendix E. National Program Evaluation Criteria *(continued)*

Criteria	Points
1. Project Summary/Approach: EPA will evaluate the quality and extent to which the narrative proposal includes a well-conceived strategy for addressing the program priorities from the statute and a well-conceived, logical strategy for achieving the anticipated environmental results associated with the proposed project in a timely manner.	25
2. Programmatic Priorities: The Agency will evaluate the quality and extent to which the proposed project addresses the programmatic priorities.	10
3. Past Performance—Programmatic Capability and Reporting on Environmental Results: The Agency will evaluate the applicant's technical ability to successfully complete and manage the proposed project, taking into account the applicant's past performance in successfully completing and managing federally funded assistance agreements.	10
4. Environmental Results—Outcomes and Outputs: The Agency will evaluate the effectiveness of the applicant's plan for tracking and measuring its progress toward achieving expected project outputs and outcomes.	15
5. Budget/Resources: The Agency will evaluate whether the proposed project budget is appropriate to accomplish the proposed goals and measurable environmental outcomes.	10
6. Clear Description of the Target Fleet: Applicants will be evaluated on the degree to which detailed information on the fleet (vessel[s], vehicle[s], and/or equipment) is provided on the Applicant Fleet Description Spreadsheet and the eligibility of technology for the fleet.	10
7. Leveraging Resources and Partnering: Applicants will be evaluated based on the extent to which they demonstrate how they will coordinate the use of EPA funding with other federal and/or nonfederal sources of funds to leverage additional resources to carry out the proposed projects and/or how that EPA funding will complement other activities.	15
8. Staff Expertise/Qualifications: Applicants will be evaluated on staff expertise/qualifications, staff knowledge, and resources or the ability to obtain them, to successfully achieve the goals of the proposed project.	5
9. Regional Significance: For the National Clean Diesel Funding Assistance Program, a factor for regional significance was added.	10

APPENDIX F: EPA Regional Clean Diesel Collaboratives

BEGINNING in 2003, EPA's regions began organizing Regional Clean Diesel Collaboratives with their states, local nonprofit organizations, private industry, and municipalities. The collaboratives listed below play a vital role in EPA's incentive-based approach to achieving superior environmental results. By tying into a network of regional stakeholders, this collaborative structure is ideal for achieving significant emission reductions across a large geographic area. Regional initiatives and state programs provide immediate and significant environmental results by working collaboratively with businesses, government and community organizations, industry, and others.

Northeast Diesel Collaborative
www.northeastdiesel.org
The Northeast Diesel Collaborative is a local initiative that builds on a foundation of voluntary action. Developed with EPA Regions 1 and 2, the initiative encourages participants to engage in projects that will reduce transportation-related air pollution to help address the high asthma rates in the Northeast. The program partners with the Northeast States for Coordinated Air Use Management (NESCAUM).

Mid-Atlantic Clean Diesel Collaborative
www.dieselmidatlantic.org
The Mid-Atlantic Clean Diesel Collaborative is a partnership among leaders from federal, state, and local governments; the private sector; and environmental groups in Delaware, Maryland, Virginia, Pennsylvania, West Virginia, and the District of Columbia.

Midwest Clean Diesel Initiative
www.epa.gov/midwestcleandiesel
EPA Region 5's Midwest Clean Diesel Initiative is a cooperative, public–private effort to reduce diesel emissions along major transportation corridors from various sectors including trucking, locomotive, construction, and ports, with an emphasis on urban areas in the Midwest.

Southeast Diesel Collaborative
www.southeastdiesel.org
The Southeast Diesel Collaborative is a voluntary, public–private partnership involving leaders from federal, state, and local governments; the private sector; and other stakeholders throughout the Southeast working to reduce diesel emissions.

Rocky Mountain Clean Diesel Collaborative

www.epa.gov/region8/air/rmcdc.html

The Rocky Mountain Clean Diesel Collaborative is a partnership of federal, state, and local governments; nonprofit organizations; the private sector; and environmental groups in Colorado, Montana, North Dakota, South Dakota, Utah, and Wyoming. The collaborative was established in November 2006.

Blue Skyways Collaborative

www.blueskyways.org

The Blue Skyways Collaborative was created to encourage voluntary air emission reduction in North America's heartland. Through partnership with nonprofit and environmental groups; private industries; and international, federal, state, and local governments, Blue Skyways strives to improve air quality.

West Coast Diesel Collaborative

www.westcoastcollaborative.org

EPA Regions 9 and 10 were instrumental in creating the West Coast Diesel Collaborative. This initiative is a partnership among leaders from federal, state, and local government; the private sector; and environmental groups in California, Oregon, Idaho, Washington, Alaska, Canada, and Mexico, committed to reducing diesel emissions along the West Coast.

Acronyms and Abbreviations

APU	Auxiliary Power Unit	NAAQS	National Ambient Air Quality Standards
CARB	California Air Resources Board	NCDC	National Clean Diesel Campaign
CCV	Closed Crankcase Ventilation		
CFS	Crankcase Filtration System	NESCAUM	Northeast States for Coordinated Air Use Management
CNG	Compressed Natural Gas	NMHC	Nonmethane Hydrocarbon
CO	Carbon Monoxide	NO_2	Nitrogen Dioxide
CO_2	Carbon Dioxide	NO_X	Nitrogen Oxides
DEQ	Diesel Emissions Quantifier	$PM_{2.5}$	Particulate Matter (Fine)
DERA	Diesel Emissions Reduction Act	PPM	Parts Per Million
DOC	Diesel Oxidation Catalyst	RFP	Request for Proposal
DPF	Diesel Particulate Filter	RIA	Regulatory Impact Analysis
EGR	Exhaust Gas Recirculation	SCR	Selective Catalytic Reduction
EPA	U.S. Environmental Protection Agency	SCRT	Selective Catalytic Reduction Technology
EPAct	Energy Policy Act	SIP	State Implementation Plan
FY	Fiscal Year	SO_X	Sulfur Oxides
HC	Hydrocarbon	ULSD	Ultra-Low-Sulfur Diesel
LNT	Lean Nitrogen Oxides Trap